T0295233

Photovoltaic Systems

Photovoltaic Systems

Artificial Intelligence–Based Fault Diagnosis and Predictive Maintenance

Edited by
K. Mohana Sundaram,
Sanjeevikumar Padmanaban,
Jens Bo Holm-Nielsen,
and
P. Pandiyan

CRC Press is an imprint of the
Taylor & Francis Group, an **informa** business

Cover Image: ME Image/Shutterstock

MATLAB® is a trademark of The MathWorks, Inc. and is used with permission. The MathWorks does not warrant the accuracy of the text or exercises in this book. This book's use or discussion of MATLAB® software or related products does not constitute endorsement or sponsorship by The MathWorks of a particular pedagogical approach or particular use of the MATLAB® software.

First edition published 2022
by CRC Press
6000 Broken Sound Parkway NW, Suite 300, Boca Raton, FL 33487-2742

and by CRC Press
2 Park Square, Milton Park, Abingdon, Oxon, OX14 4RN

© 2022 selection and editorial matter, K. Mohana Sundaram, Sanjeevikumar Padmanaban, Jens Bo Holm-Nielsen, and P. Pandiyan; individual chapters, the contributors

CRC Press is an imprint of Taylor & Francis Group, LLC

Reasonable efforts have been made to publish reliable data and information, but the author and publisher cannot assume responsibility for the validity of all materials or the consequences of their use. The authors and publishers have attempted to trace the copyright holders of all material reproduced in this publication and apologize to copyright holders if permission to publish in this form has not been obtained. If any copyright material has not been acknowledged please write and let us know so we may rectify in any future reprint.

Except as permitted under U.S. Copyright Law, no part of this book may be reprinted, reproduced, transmitted, or utilized in any form by any electronic, mechanical, or other means, now known or hereafter invented, including photocopying, microfilming, and recording, or in any information storage or retrieval system, without written permission from the publishers.

For permission to photocopy or use material electronically from this work, access www.copyright.com or contact the Copyright Clearance Center, Inc. (CCC), 222 Rosewood Drive, Danvers, MA 01923, 978-750-8400. For works that are not available on CCC please contact mpkbookspermissions@tandf.co.uk

Trademark notice: Product or corporate names may be trademarks or registered trademarks and are used only for identification and explanation without intent to infringe.

Library of Congress Cataloging in Publication Data
A catalog record has been requested for this book

ISBN: 978-1-032-06426-0 (hbk)
ISBN: 978-1-032-06428-4 (pbk)
ISBN: 978-1-003-20228-8 (ebk)

DOI: 10.1201/9781003202288

Typeset in Times
by SPi Technologies India Pvt Ltd (Straive)

Contents

About the Editors

Dr. K. Mohana Sundaram received a BE in electrical and electronics engineering from University of Madras in 2000, a MTech in high voltage engineering from SASTRA University in 2002, and a PhD degree from Anna University, India in 2014. His research interests include intelligent controllers, power systems, embedded system, and power electronics. He has completed a Rs. 3.1 million project sponsored by DST, government of India. Currently he is working as a professor in the EEE department at KPR Institute of Engineering and Technology, India. He has produced four PhD candidates under his supervision in Anna University, Chennai. He has published three books and is serving as a reviewer for IEEE, Springer, and Elsevier journals. He is an active member of IE, ISTE, and IAENG. He has published close to 60 papers in international journals.

Sanjeevikumar Padmanaban (Senior Member, IEEE) received the degree in electrical engineering from the University of Bologna, Bologna, Italy, in 2012. He was an associate professor with VIT University from 2012 to 2013. In 2013, he joined the National Institute of Technology, India, as a faculty member. In 2014, he was invited as a visiting researcher with the Department of Electrical Engineering, Qatar University, Doha, Qatar, funded by the Qatar National Research Foundation (Government of Qatar). He continued his research activities with the Dublin Institute of Technology, Dublin, Ireland, in 2014. He has served as an associate professor with the Department of Electrical and Electronics Engineering, University of Johannesburg, Johannesburg, South Africa, from 2016 to 2018. From March 2018 to February 2021, he was a faculty member with the Department of Energy Technology, Aalborg University, Esbjerg Campus, Denmark. He is currently working as an associate professor with the CTIF Global Capsule (CGC) Laboratory, Department of Business Development and Technology, Aarhus University, Herning, Denmark. He is a Fellow of the Institution of Engineers, India, the Institution of Electronics and Telecommunication Engineers, India, and the Institution of Engineering and Technology, UK. He was a recipient of the Best Paper cum Most Excellence Research Paper Award from IET-SEISCON'13, IET-CEAT'16, IEEE-EECSI'19, IEEE-CENCON'19, and five best paper awards from ETAEERE'16 sponsored Lecture Notes in electrical engineering, Springer book. He is also an editor/associate editor/editorial board member of refereed journals, in particular the *IEEE Systems*, *IEEE Transactions on Industry Applications*, *IEEE Access*, *IET Power Electronics*, *IET Electronics Letters*, and *Wiley-International Transactions on Electrical Energy Systems*; a subject editorial board member of *Energy Sources—Energies Journal*, MDPI; and the subject editor of the *IET Renewable Power Generation, IET Generation, Transmission and Distribution, and FACTS Journal* (Canada).

Jens Bo Holm-Nielsen, PhD was born in 1954. He is currently the head of the Research Group of Bioenergy and Green Engineering, Department of Energy Technology, Aalborg University, Denmark. He has 30 years of experience in the field of biomass feedstock production, biorefinery concepts, and biogas production. He was a board member of research and development committees of the cross-governmental body of biogas developments, Denmark, from 1993 to 2009. He is also a secretary and/or chair of several biogas and bioenergy nongovernmental organizations. He is also Chair and Presenter of Sustainable and 100 percent Renewables and SDG-17 goals. He has experience in a variety of EU projects, and is an organizer of international conferences, workshops, and training programs in EU, USA, Canada, China, Brazil, India, Iran, Russia, and Ukraine, among others. His research interests include managing research, development, and demonstration programs in integrated agriculture, environment, and energy systems. He has fulfilled the biomass and bio-energy research and development projects. His research interests also include biofuels, biogas, and biomass resources. He conducts international courses, training programs, and supervision for MSc and PhD students and academic staff, governmental bodies, and experts in bioenergy systems.

Dr. P. Pandiyan received his Bachelor's in electrical and electronics engineering from Anna University, Chennai, India, in 2006; Master's (M.Tech – VLSI Design) in electronics and communication engineering from Sathyabama University, Chennai, India, in 2010; and PhD in instrumentation and control engineering from the National Institute of Technology, Tiruchirappalli, India, in 2018. His PhD work is in the area of MEMS-based logic devices. His research interests include design and simulation of MEMS-based logic devices, energy harvesting, and electric vehicles. He has received a Rs. 500,000 grant from AICTE toward the conduction of STTP. He has seven years of teaching experience at various engineering colleges and three years of research experience at NIT, Trichy. Currently he is an associate professor in EEE department at KPR Institute of Engineering and Technology, India. He has organized various value-added courses, workshops, seminars, and guest lectures. A PhD scholar is working under his supervision in the area of hybrid energy harvesting. He has published 15 papers in reputed international journals and 15 papers in the proceedings of national and international conferences.

List of Contributors

Johny Renoald Albert
Assistant Professor, Department of Electrical and Electronics Engineering, Vivekanandha College of Engineering for Women, Namakkal, Tamilnadu, India

T. Chinnadurai
Department of ICE, Sri Krishna College of Technology, Coimbatore, Tamil Nadu, India

A. Gayathri
Department of EEE, Sri Krishna College of Technology, Coimbatore, Tamil Nadu, India

P. Jamuna
Department of Electrical and Electronics Engineering, Nandha Engineering College, Erode, Tamilnadu, India

G. Sophia Jasmine
Department of Electrical and Electronics Engineering, Sri Krishna College of Technology, Coimbatore, Tamil Nadu, India

N. Yogambal Jayalashmi
Department of Electrical and Electronics Engineering, Dr. Mahalingam College of Engineering and Technology, Pollachi, India

Thenmalar Kaliannan
Professor, Department of Electrical and Electronics Engineering, Vivekanandha College of Engineering for Women, Namakkal, Tamilnadu, India

Selvakumar Kuppusamy
Assistant Professor, Department of EEE, SRM Institute of Science and Technology, Chennai, Tamilnadu, India

K. Lakshmi
Department of Electrical and Electronics Engineering, Sri Krishna College of Technology, Coimbatore, Tamil Nadu, India

D. Magdalin Mary
Department of Electrical and Electronics Engineering, Sri Krishna College of Technology, Coimbatore, Tamil Nadu, India

V. Manimegalai
Department of EEE, Sri Krishna College of Technology, Coimbatore, Tamil Nadu, India

P. Pandiyan
Department of EEE, KPR Institute of Engineering and Technology, Coimbatore, Tamil Nadu, India

Madhumathi Periasamy
Research scholar, Full-time, Department of Electrical and Electronics Engineering, Anna University, Chennai, Tamil Nadu, India

N. Prabaharan
School of Electrical & Electronics Engineering, SASTRA Deemed University, Thanjavur, Tamil Nadu, India

T. Rajasekaran
Department of CSE, KPR Institute of Engineering and Technology, Coimbatore, Tamil Nadu, India

V. Rukkumani
Department of EIE, Sri Ramakrishna Engineering College, Coimbatore, Tamil Nadu, India

S. Saravanan
Department of Electrical and Electronics Engineering, Sri Krishna College of Technology, Coimbatore, Tamil Nadu, India

Fantin Irudaya Raj Edward Sehar
Assistant Professor, Department of Electrical and Electronics Engineering, Dr. Sivanthi Aditanar College of Engineering, Tiruchendur, Tamilnadu, India

R. Senthilkumar
Department of Electrical and Electronics Engineering, Sri Krishna College of Technology, Coimbatore, Tamil Nadu, India

Gopinath Singaram
Associate Professor, Department of Electrical Engineering, Annasaheb Dange College of Engineering and Technology, Sangli, Maharashtra, India

K.P. Suresh
Department of Electrical and Electronics Engineering, Sri Krishna College of Technology, Coimbatore, Tamil Nadu, India

M. Suresh
Department of Electronics and Communication Engineering, Kongu Engineering College, Erode, Tamil Nadu, India

Ramji Tiwari
Department of EEE, Sri Krishna College of Engineering and Technology, Coimbatore, Tamil Nadu, India

K. Umamaheswari
Department of Electrical and Electronics Engineering, Dr. Mahalingam College of Engineering and Technology, Pollachi, India

T. Yuvaraj
Department of EEE, Saveetha School of Engineering, Saveetha Institute of Medical and Technical Sciences, Saveetha University, Chennai, Tamil Nadu, India

1 Online Fault Diagnosis and Fault State Classification Methods for PV Systems

V. Manimegalai and A. Gayathri
Sri Krishna College of Technology, Coimbatore, India

P. Pandiyan
KPR Institute of Engineering and Technology, Coimbatore, India

V. Rukkumani
Sri Ramakrishna Engineering College, Coimbatore, India

CONTENTS

DOI: 10.1201/9781003202288-1

LEARNING OUTCOME

i. To understand the importance of online fault diagnosis and fault state classification methods in solar photovoltaic (PV) systems
ii. To apply artificial intelligence techniques for detecting online fault diagnosis
iii. To apply artificial intelligence techniques for fault state classification using performance index of the solar PV systems

1.1 INTRODUCTION

Today, we are living in a world where everything hinges on electricity as a major source of energy for everyday life and for small and large-scale industries. This increases the demand for electricity, which today is at an all-time high. In India, total electricity consumption for 2019–2020 was 1,291,494 GWh, with agriculture accounting for 17.7%, domestic accounting for 24%, industrial accounting for 8%, and transportation and railways accounting for 1.5% and 6.1%, respectively. Figure 1.1 shows that electricity consumption in 2019–2020 increased compared to the previous year, indicating that electricity utilization has increased as a result of technological advancement. If consumption rises, demand rises, and so does the amount of fossil fuel required for production.

As per 2021 statistics, demand for all fossil fuels is expected to grow substantially. Worldwide, demand for coal is expected to rise by 4.5%, natural gas demand is expected to rise by 3.2%, electricity demand is expected to rise by 4.5%, and demand for renewable energy is expected to rise by 3% across all sectors, including power and industry. The electricity sector is at the forefront, with demand for renewable energy at more than 8% and rising.

The installed capacity of renewable electricity generation sources in 2020, excluding hydro from utilities, increased by 12% over 2019. The number of thermal sources

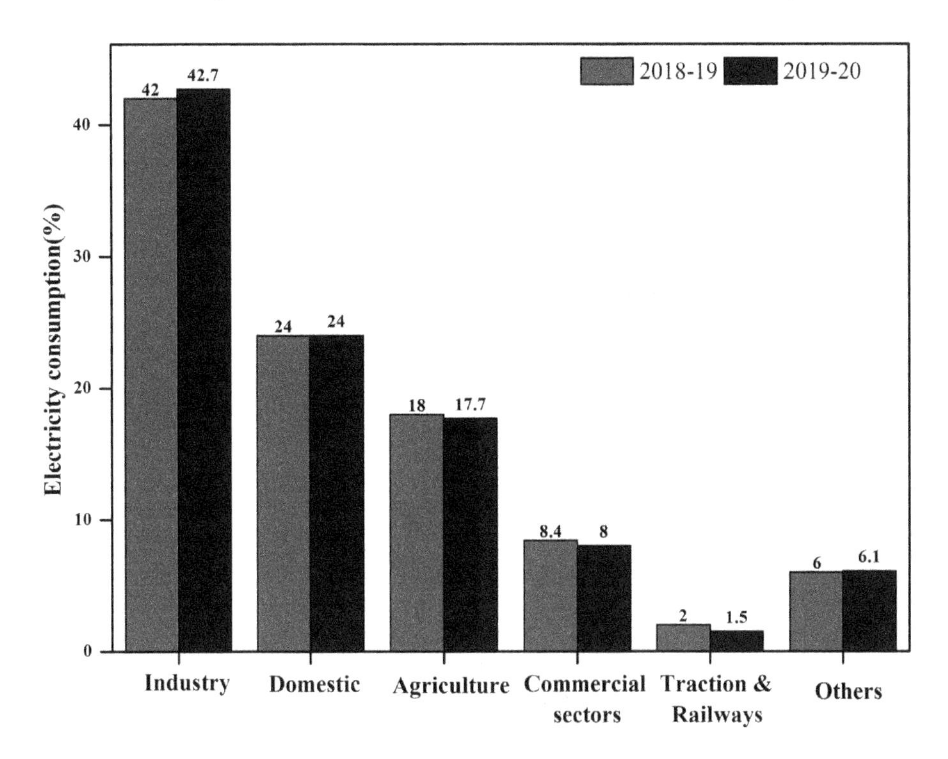

FIGURE 1.1 Sector-wise electricity consumption.

increased by only 1.91%. This means that in the coming years, renewable energy will play a significant role in the generation of electricity. Solar and wind, in particular, are expected to supply two-thirds of renewable growth [1]. The share of renewable in the power sector era is anticipated to extend to nearly 30% in 2021. Wind is on track to grow at the fastest rate in the renewable era, increasing by 275 TWh, or roughly 17%, from 2020.The Sun-based PV power production is anticipated to rise by 145 TWh, or nearly 18%, and to approach 1000 TWh in 2021.

Rising concern about climate change, the effects of air pollution, energy security, oil prices, and the alternate low-carbon technology options has created a strong impetus to use renewable energy. As per India's statistics for 2021, the high latent capacity for the generation of renewable energy from different sources such as wind, solar, biomass, small hydro, and cogeneration bagasse is shown in Figure 1.2. The total power generated by renewable sources in the country as of 2020 is estimated at 1,097,465 MW. Solar generates 748,990 MW (68.25%), wind generates 302,251 MW (27.54%), small hydro generates 21,134 MW (1.93%) with a 100-foot hub, biomass generates 17,536 MW (1.60%), bagasse-based cogeneration in sugar mills generates 5,000 MW (0.46%), and other waste generates 2,554MW (0.23%). According to Figure 1.2, the solar PV system plays a significant role in all renewable energy sources. Solar power will clearly continue to be an essential renewable option in the coming decades. Electricity generated by solar photovoltaic is critical for reducing demand.

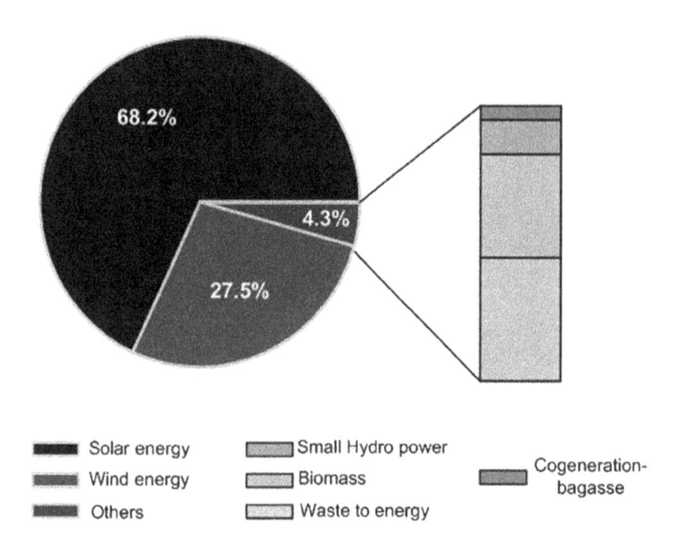

FIGURE 1.2 Generation of renewable with various sources.

Even though the solar cell is clean and free from pollution, it is still difficult to meet the demand directly. Therefore, it produces a variable output based on area and on irradiance and temperature. In order to capture maximum energy from a photovoltaic system, various technologies have been used. However, the efficiency of energy production from solar photovoltaic cells has decreased as a result of greater challenges such as climate change, dust depositions, partial shading, and bird waste, as well as faults caused by internal or external factors. As a result, both the efficiency and life span of solar cells have decreased. Therefore, it is critical to monitor the solar photovoltaic cell, and if a fault occurs, an alert signal should be generated in accordance with technological trends, and it should be used to determine the nature of the fault and classify those faults in order to bring up the appropriate algorithm to sort out the problem as soon as possible based on the type of fault and failure of the PV system.

Maintenance engineers seek to develop economical, efficient, dependable methods and tools to monitor and rapidly identify fault states, cut maintenance planning time, and minimize energy consumption. The most important parameter to identify the solar fault state and reduction in output power is based on performance index (PR) [2–7]. This index also helps evaluate the solar power plant's performance and aging. If a PV power plant experiences a fault, an operator can determine its extent by calculating the PR index, which is less than the normal value. Then the operators can identify the fault and work to rectify it. If the annual PR index value is not constant, the operator cannot identify the faults because of changes in temperature, which affect the output power characteristics. The change in temperature of a solar plant can be identified by Weather-Corrected Performance Ratio (WCPR) [8]. PV power plants with WCPR are corrected for temperature distortion by adjusting power output based on the average solar cell temperature. There are certain methods to find the different fault states, such neural networks, fuzzy logic, different AI techniques [9], and other techniques for detecting outliers in photovoltaic power plants.

The rest of this chapter is structured as follows: Sections 1.2 and 1.3 deal with types of faults and classifications used to identify the location of PV systems. Comparison of common faults in PV systems, with prevention techniques, can be found in Section 1.4. Fault detection and diagnosis using electrical techniques are discussed in Section 1.5. Section 1.6 presents the classification of fault states for solar power plants to find out the output power reduction. In Section 1.7, the evaluation of PV plant based on clearness index and proposed variable index is discussed. The different techniques for detecting outliers in PV plants for better output are presented in Section 1.8, which concludes that the online fault diagnosis method is an efficient method to identify the faults in large solar plants.

1.2 TYPES OF FAULTS AND ITS CLASSIFICATIONS

The fault is the deviation of voltages and currents from nominal values or states. When a fault occurs, it causes excessively high currents to flow, which causes damage to equipment and devices. The faults are mainly of two types: open circuit and short circuit. They are further classified as symmetrical or unsymmetrical.

1.2.1 Open Circuit Faults

Open-circuit faults are most commonly caused by the failure of one or more of the conductors used in the electrical system. Faults in overhead wires and cables, as well as failures in the phase of circuit breakers, conductor melting, and fuse failure, are all possible causes of this issue. This fault is also called a series fault, which is shown in Figure 1.3. This type of fault causes the voltage to exceed the normal value and causes insulation failures and can potentially develop into a short circuit. These faults are classified as follows:

1. Open conductor fault
2. Two conductors open fault
3. Three conductors open fault

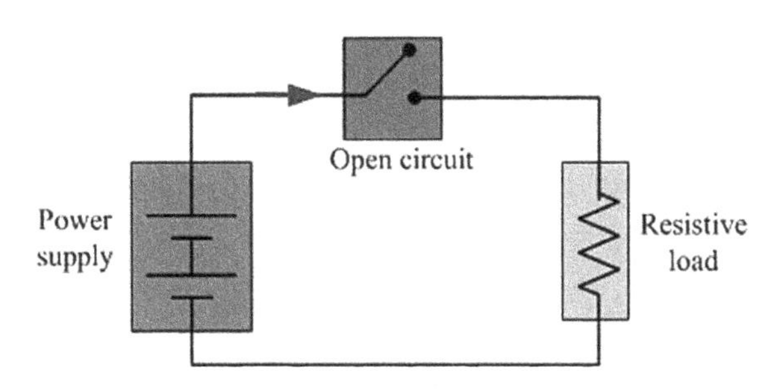

FIGURE 1.3 Open-circuit fault.

1.2.2 Short-Circuit Faults

Short-circuit faults are most commonly caused by a failure in the insulation between the phase conductors and the earthing conductor. These are the most prevalent forms of faults, and they are characterized by an abnormally high current flow throughout the transmission lines or other equipment. Even if a short-circuit failure only lasts for a brief period, it can be extremely destructive to the apparatus. This fault is also called a shunt fault, which is illustrated in Figure 1.4. These faults occur due to internal or external factors, damaged equipment, improper installation, lightning-caused surges, etc. These faults cause abnormal current flow that leads to overheating the equipment, which reduces the life span of the installation.

1.2.3 Symmetrical Faults

Symmetrical faults are serious defects that rarely occur in power systems. These types of defects are also known as balanced faults. If this type of fault occurs, the system remains balanced but serious damages are caused to the electrical power system equipment. This type of defect is easy to analyze. They are classified as line to line to ground (L-L-L-G) or line to line (L-L-L) fault, as shown in Figure 1.5.

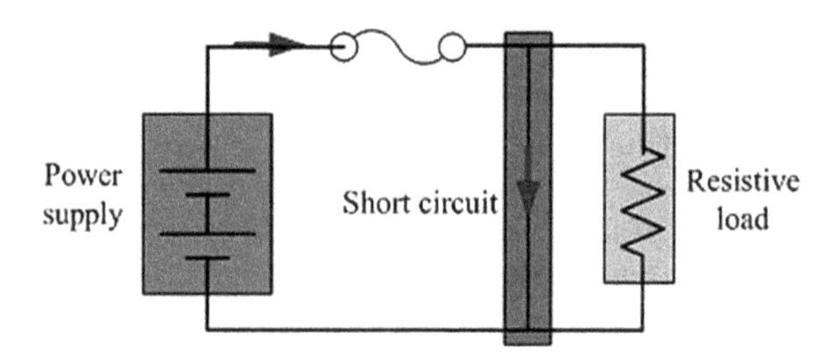

FIGURE 1.4 Short-circuit fault.

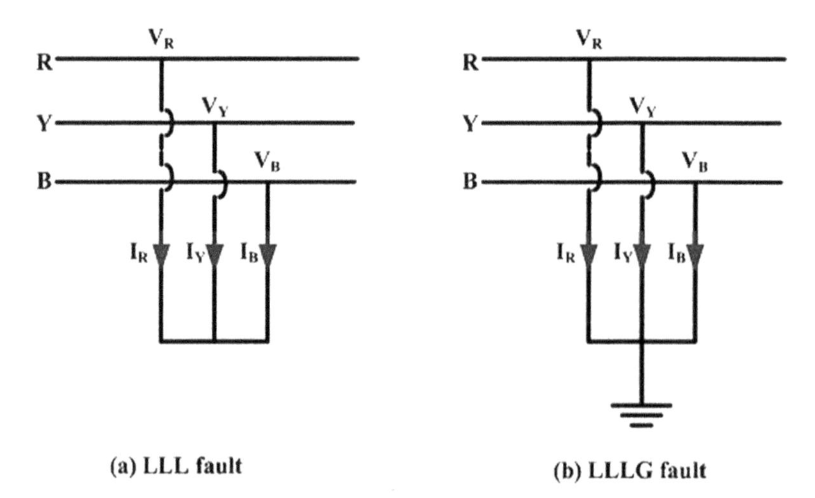

FIGURE 1.5 LLL fault and LLLG fault.

1.2.4 Unsymmetrical Faults

Unsymmetrical faults tend to be less serious than symmetrical faults. Figure 1.6 depicts the classification of line-to-ground (L-G), line-to-line (L-L), and double line–to-ground (LL-G) faults. Line-to-ground faults are the most common type of fault and occur when a conductor comes in contact with the ground. Line-to-line faults occur when two conductors come in contact with each other. Line-to-line faults are also referred to as unbalanced faults because they create an imbalance in the system.

1.3 FAULTS IN PV SYSTEMS

The photovoltaic cell absorbs energy from the sun and converts it into electricity [10]. It works on the principle of photoelectric effect. The main part of the PV cell is made of silicon, which is a good semiconductor. When photons radiating from the sun strike the silicon, they loosen the electron from a silicon atom. The proper imbalance is required to convert this free electron into electric current.

A single solar cell produces very little electricity and can fulfill no load on the application side. Furthermore, the type of cell utilized in the manufacturing process might influence the efficiency by up to 25%. Thus, in order to meet the increased power requirement, solar PV cells must be interconnected as groups of cells in a series or parallel arrangement. Such a collection of cells is called a solar photovoltaic module, and the collection of modules is a solar photovoltaic array. Figure 1.7 illustrates the photovoltaic cell, array, and module. The solar modules can be of different type, such as monocrystalline, polycrystalline, or amorphous.

In solar PV systems, hot spots, delamination, discoloration, and corrosion are common problems caused by the salty and damp atmosphere found around the coasts and in high-temperature zones, among other factors.

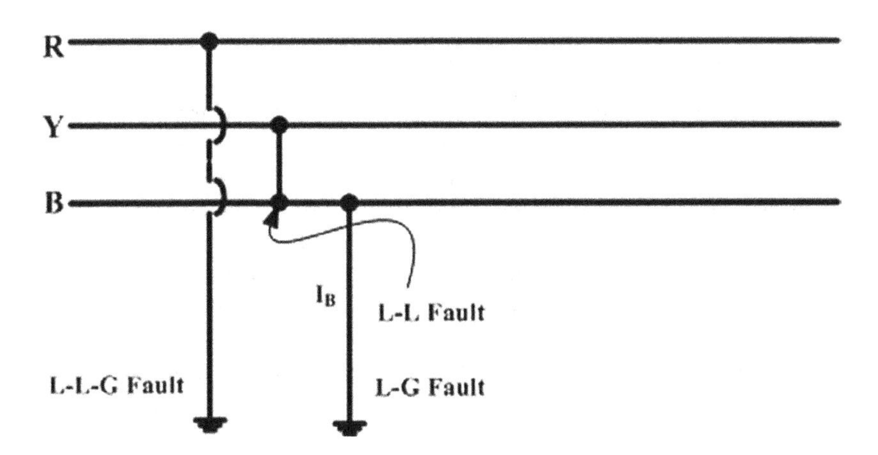

FIGURE 1.6 Line-to-ground fault, line-to-line fault, double line–to-ground fault.

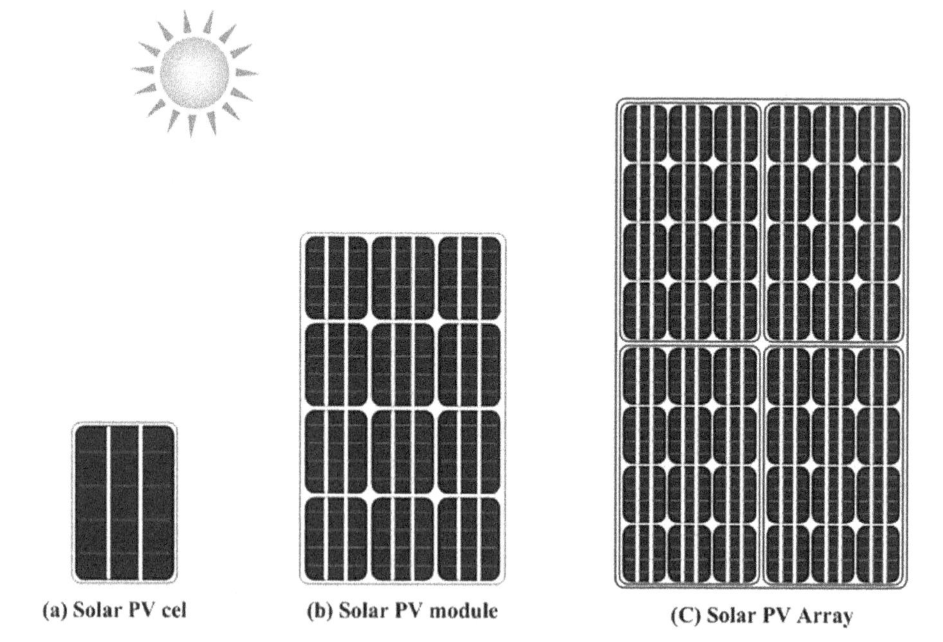

FIGURE 1.7 Solar cell, solar module, solar array.

1.3.1 Hot Spots in PV Module

A hot spot shown in Figure 1.8 occurs within the solar PV module when more cells get shaded while connected with unshaded cells in a string. Once the number of shaded cells exceeds the operating current of a module, it leads to a short circuit, which may result in breakage of glass due to exposure of heat in a particular spot [11].

1.3.2 Delamination of PV Module

Another factor for causing corrosion is delamination due to moisture and water ingress [10], illustrated in Figure 1.9. In the solar PV module, the cell is coated with ethylene vinyl acetate, which protects the cell from dirt, moisture, temperature changes, and UV radiation. During manufacturing, cell coating takes place at specified temperature and pressure conditions. If the process is not performed properly, then the ethylene vinyl acetate coating starts to dissolve fairly quickly when exposed to varying weather conditions, and mixes with moisture to form a white film [10]. This film affects both the performance and life span of the cell. This type of corrosion can be minimized by providing proper drainage within the module to avoid water accumulation, as well as by using infrared thermography techniques [12].

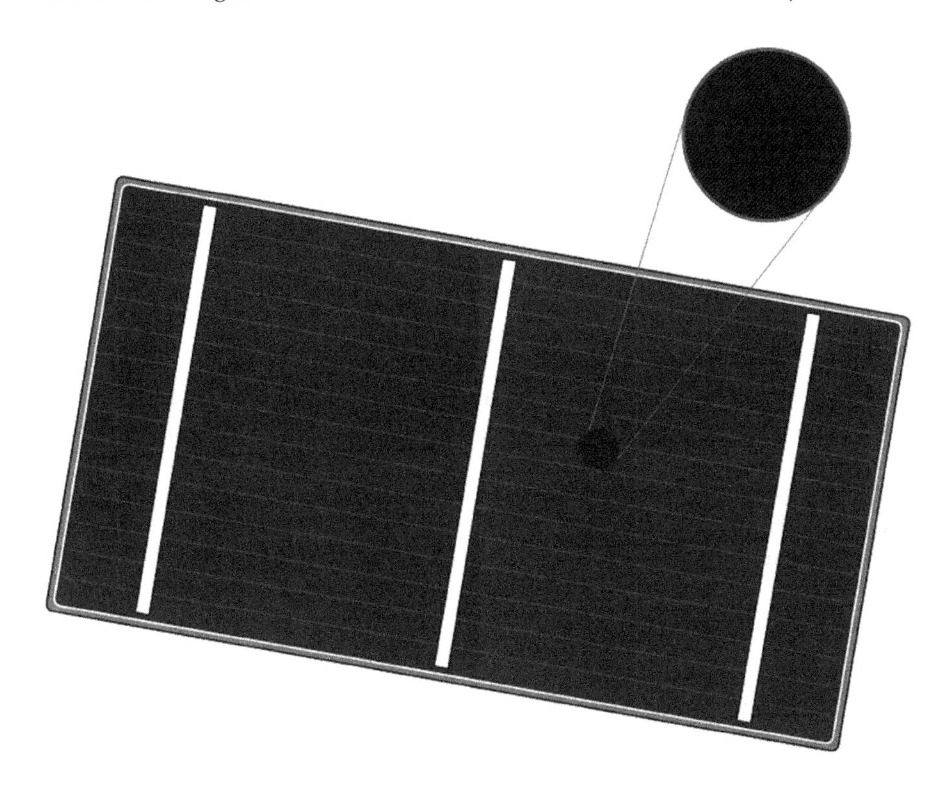

FIGURE 1.8 Hot spot in PV module.

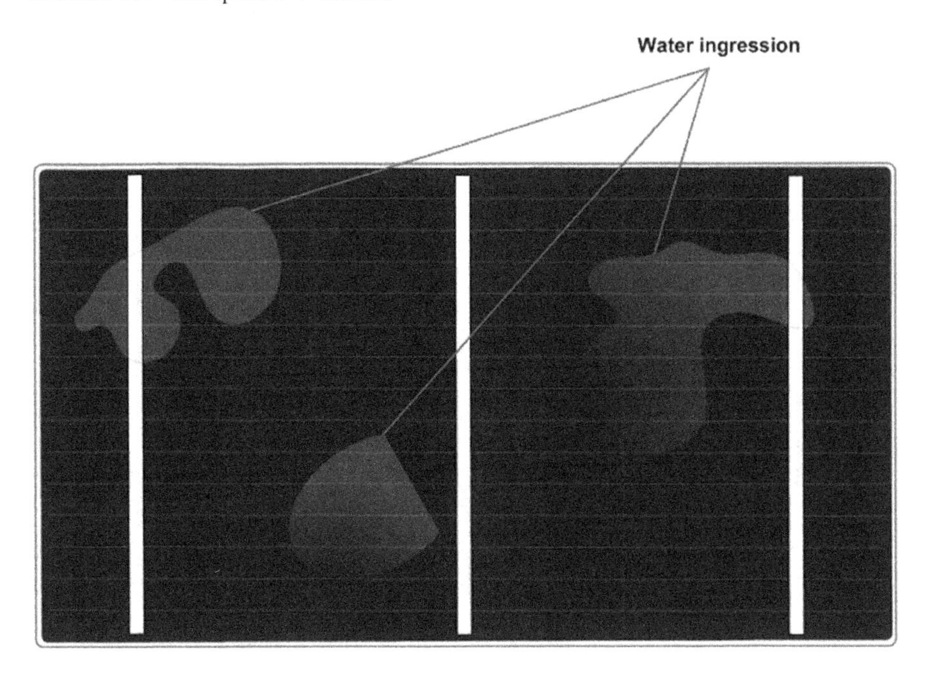

FIGURE 1.9 Delamination in PV array.

1.3.3 Discoloration of Encapsulation of PV Module

Encapsulation is simply the exposed top surface of the solar PV module. It represents the mechanical strength of the module and guards the solar cell. It functions as a barrier between the solar PV cell and the glass enclosure. This discoloration occurs due to aging of PV modules and operation of PV modules for a long period of time. This type of fault occurs in areas of high heat and low humidity, and it is shown in Figure 1.10.

1.3.4 Corrosion

Corrosion in the PV system usually occurs in a string of cells due to a glass made of tin oxide when operated at high voltage. It is possible to reduce corrosion in the PV module encapsulant by decreasing the ionic conductivity of the material. In some instances, corrosion might result in a loss of power, which can result in the complete failure of the system. When using crystalline PV modules, corrosion of the front contact is influenced by the metallization as well as the encapsulant system [12]. With the exception of the glass, it can be found on metal and other regions of the module, as indicated in Figure 1.11.

1.3.5 Micro Crack, Snail Trail, and Cell Damage

The PV panel should be handled with proper care by the installer during the installation process. If any load is kept on the top of the panel and if it is transported from one place to another without proper care, the panel may crack, which affects the performance of the cell [13]. This damaged cell produces heat quickly and leads to hot spots on the cell, which reduces the efficiency of the cell. Other environmental conditions such as inclement weather, blown dust and debris, and animal secretions

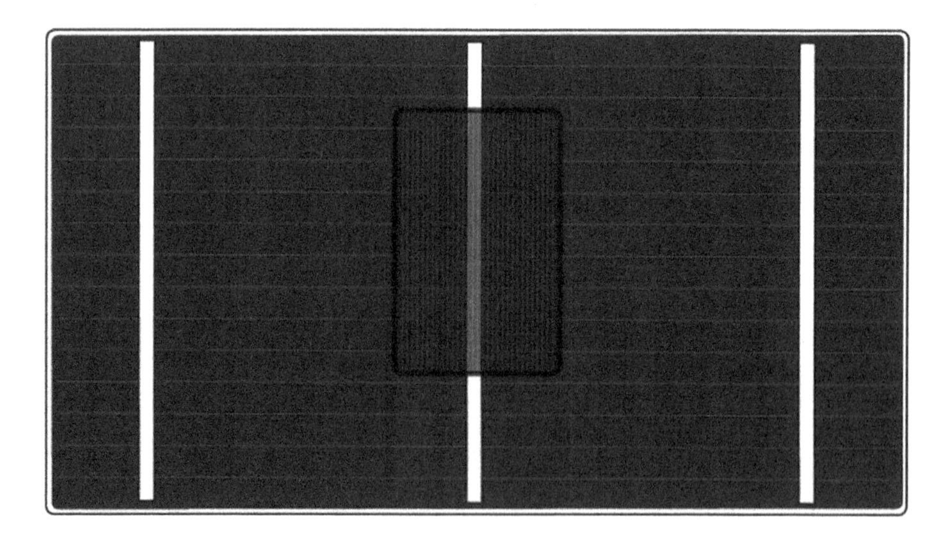

FIGURE 1.10 Discoloration of PV encapsulation.

FIGURE 1.11 Corrosion in string of PV cell.

(snail trails) may also lead to micro cracks in a PV system, as shown in Figure 1.12. Due to these cracks, the dust particles and moisture enter into the encapsulant and lead to delamination, discoloration, and corrosion.

1.3.6 Shading Fault

When the solar panel gets sufficient sunlight, more voltage is produced and stored in a battery. If the solar panel is shaded from the sun, the voltage produced is zero, and then the voltage stored in the battery gets discharged by moving the current in the opposite direction. In such instances, the bypass diode prevents the flow of current in the opposite direction and also prevents the battery from discharging, as illustrated in Figure 1.13. If there is a failure of the bypass diode, a fault like hot spot occurs in the module, which sometimes results in a short circuit or fire in the PV module [14].

1.4 COMPARISON OF FAULTS

The comparison of common faults in the PV system with some prevention techniques is provided in Table 1.1.

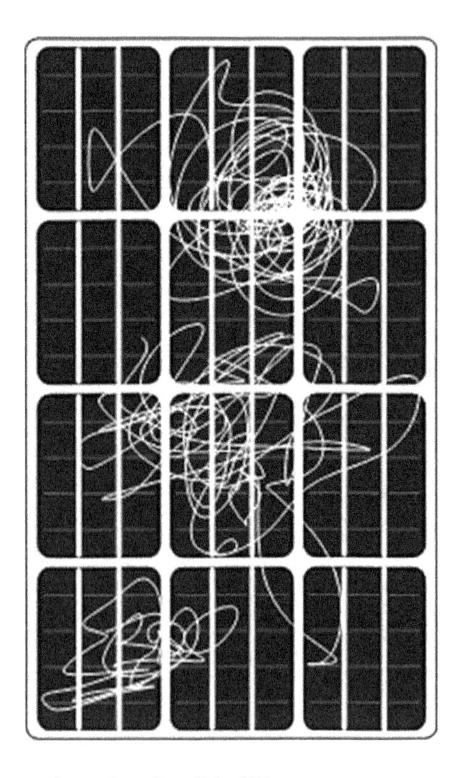

FIGURE 1.12 Micro cracks and snail trail in PV.

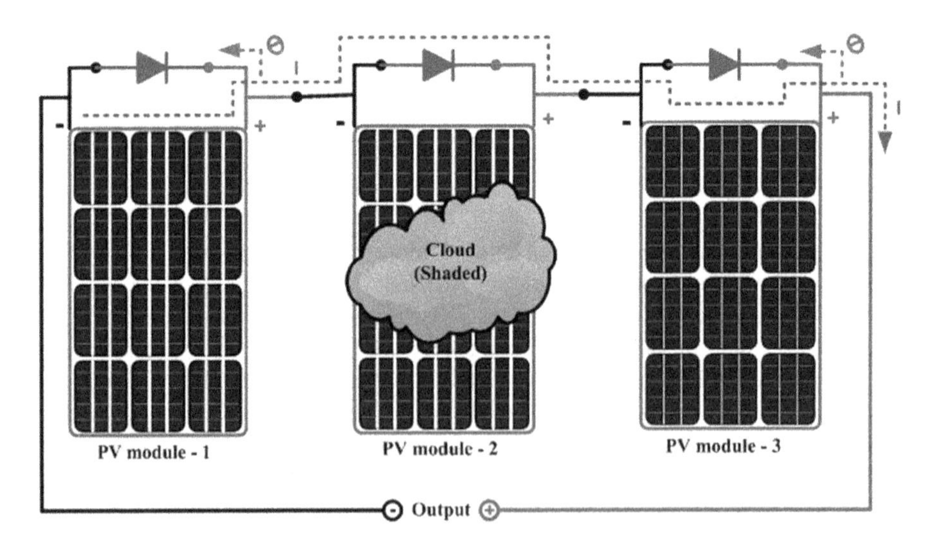

FIGURE 1.13 Shaded PV is protected by bypass diode.

TABLE 1.1
Comparison of Common Faults in a PV System

S.No	Type of Fault	Location	Occurrence of Fault	Prevention Technique
1.	Line-line fault	Appears in PV array/ module	When the two points get short-circuited	This fault is identified by over-current protection device (OCPD) [15]
2.	Ground fault	Appears in PV array/ module[16]	i) This is due to insulation failure of cables/ conductors ii) The short circuit occurred between conductors and ground	Mersen provides PV-rated fuses for all ground-fault protection circuits
3.	Hot spots	Appears in PV cells or PV modules	It may appear due to overheating of cell, which leads to high resistance and partial shading of cell	Infrared thermography along with some other algorithms is used to clear hot spots
4.	PV module fault	Appears in PV module	Cracks on the glass, disconnection of cell	Identified by Earth Capacitance Measurement (ECM) method and thermography camera [17]
5.	Shading fault	Appears in PV panels or PV module	Due to covering of panel by foliage and bird/insect droppings, as well as falling within a shadow of a tall building or tower	This is identified by sensors and provides suitable cleaning methods for bird waste, leaves, and mud, but there is no effective fix for building and tower shadowing [18]
6.	Bypass diode fault	Diode on PV module	Due to overheating of cell	Avoid continued shading of panel for a longer period. No effective fixes [19]
7.	Arc fault	Appears in PV array	When cells get overheated, plasma gets discharged and sometimes burns the array	An arc fault circuit interrupter is a standard protection device [20]
8.	Junction box fault	Photovoltaic system	Energy loss stress from system, cable without welding, improper rework connection	This fault is prevented by avoiding system overstress, monitoring proper installation of cables and connections [21]

1.5 DETECTION AND DIAGNOSIS OF FAULTS IN PV

The monitoring of solar panels is the only way to detect the different faults that occur in PV modules. The different techniques are used to detect and diagnose the defects in the PV system. There are three main classifications of fault-diagnosing techniques: visual, thermal, and electrical. In this chapter, the electrical technique is briefly explained with further subclassification.

The electrical techniques are subclassified into:

1) Voltage and current measurement
2) IV characteristics analysis
3) AI method
4) Statistical and signal processing
5) Based on power losses

1.5.1 Voltage and Current Measurement

This technique is used to analyze the fault in a PV array by considering the current and voltage parameter. This fault is categorized by determining the output current and variation of voltages in each branch of the system. If the output current in the branch is reduced to less than 10% of rated current, then it results in no faults. If the output current at the desired branch is zero, it results in an open-circuit fault. If the current across the branch is less than the rated current but the voltage value is greater than the desired value, then it results in a short circuit. If the drop in output current is more than 40% of rated current, then it is considered a hot spot in the PV module. In this case, a suitable and easy fault detection method should be used to identify the type of fault. This type of analysis is used to decrease the losses in the solar power plant [22].

1.5.2 IV Characteristics Analysis

This method is based on comparing the threshold value with actual values obtained from the current-voltage characteristics of the PV system. By performing this technique, any disconnection in the array can short out easily. If a fault occurs in the array, this analysis can be implemented in a fault protective device. If faults like a hot spot or a diode fault occur in a PV system, this analysis is easy to implement for a medium-scale PV array. If the fault occurred in the module, the fault can be easily diagnosed at medium cost [23, 24].

1.5.3 Fault Diagnosis Based on AI Method

This method is used to find different types of faults that occur in PV systems, like faults in power devices, overheating, and lesser output voltage. The voltage fault that occurred in the PV system can be detected using the Takagi Sugeno Kahn Fuzzy Rule & Bayesian belief network [25, 26]. This approach is analogous to the voltage and current measurement method, but it compares the actual output power with measured power. If the variation of power occurs above the threshold value, then a fault can be detected. The two fault diagnosis algorithms such as hybrid support vector machine and k-nearest neighbor algorithm are used to detect a shorted fault and fault occurring in a blocking and bypass diode. By using this method, a fault that occurs in the inverter can be shorted out by VI characteristics. The Artificial Neural Network method (ANN) is used in a PV system to verify the entire power generation system operation with expected output production with necessary parameters like peak voltage, short circuit current, and peak power with VI characteristics. This method of fault diagnosis is accurate and reliable.

1.5.4 Statistical and Signal Processing–Based Approach

The three main methods used in this approach are Time Domain Reflectometry (TDR), Earth Capacitance Measurement (ECM), and Speared Spectrum TDR (SSTDR), which are used to identify and position the faults in solar photovoltaic modules [17]. Time Domain Reflectometry is to detect and localize the failed PV modules, but it can be easily affected by installation conditions [27]. The next method is ECM, which is used to detect which module is disconnected from the string based

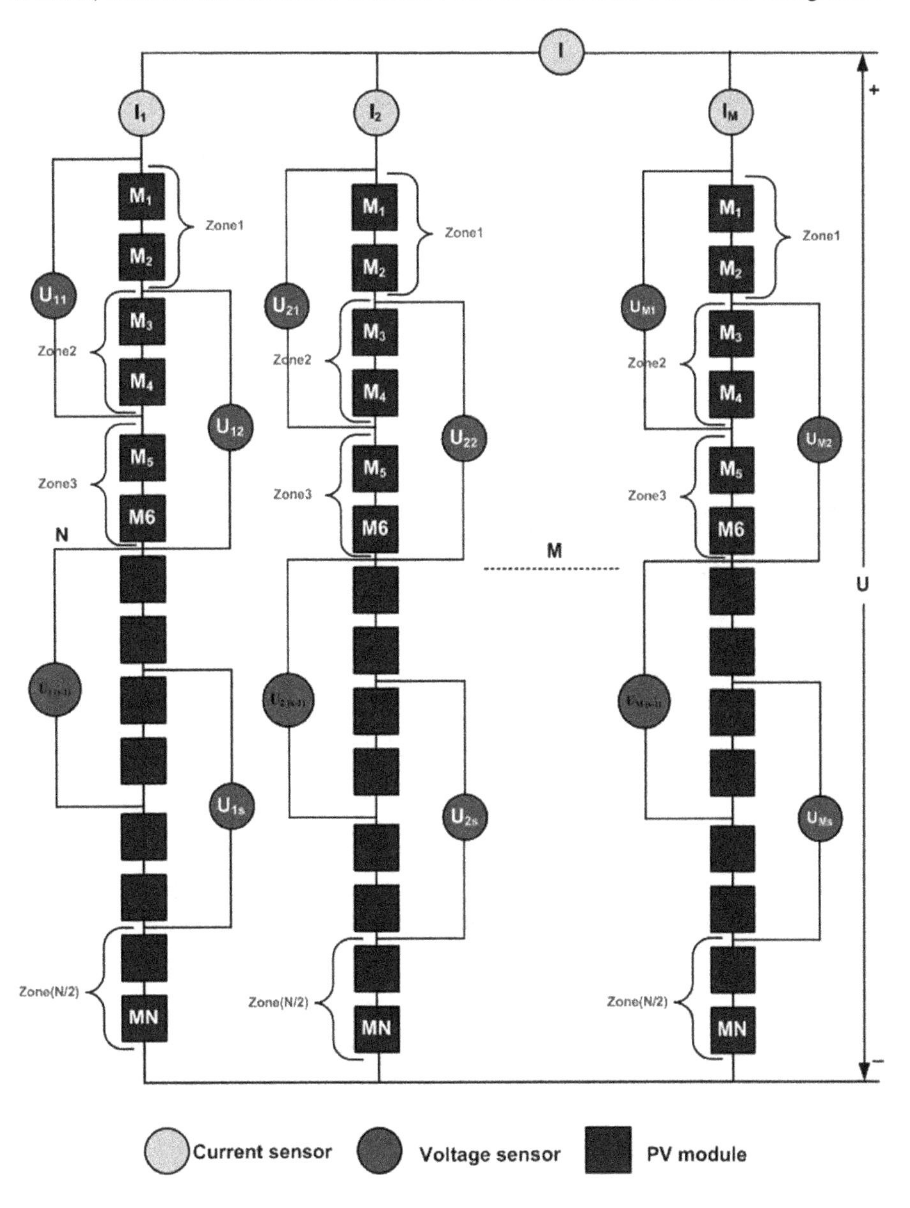

FIGURE 1.14 Solar PV array with voltage and current detection method.

on measurement without the effect of the irradiance variation, and the SSTDR can identify the location of the degradation of cell like the increase in the resistance between the PVs [28]. This method is used to detect and diagnose complex faults at an economical cost.

1.5.5 Methods Based on Power Losses

This method is used to analyze the module of a string, a fault in a string of a cell, aging of a panel, MPPT error, partial shading error, etc. [29]. It diagnoses the fault that occurs due to object shading or small settlement of dirt, and any losses occurring in the cable are determined experimentally. Multiple faults can be detected with the help of the automatic supervision algorithm by identifying two parameters, such as thermal and miscellaneous capture losses, to find the type and location of fault in the PV system. These data are collected and compared with the threshold value to identify the type of fault. This method is not prominent for identifying the fault that occurs on the AC side (Figure 1.14).

1.6 CLASSIFICATION OF FAULT STATES IN SOLAR PV SYSTEMS

1.6.1 Performance Index of Solar PV System

1.6.1.1 Performance Ratio (PR)

In accordance with the IEC 61724-1 standard [2], a performance indicator for solar power plants – called PR – has been developed. This index represents the overall efficiency of a solar PV power plant. In the event of a solar power plant fault, a PR index that has a lower value can be used to identify the fault state. An application of these indices can help design and maintain a solar PV power plant. In addition, this index has the ability to identify and classify fault states and aging faults, which becomes important when making asset management decisions.

$$PR = \frac{\sum_{t=1}^{24} P_{AC}(t)}{\sum_{t=1}^{24} G(t) A\eta} \tag{1.1}$$

where P_{AC} represents the power output (AC) of the solar PV power plant, A denotes the area occupied by the solar plant, G corresponds to the irradiation, and efficiency of the solar PV plant represented by η.

1.6.1.2 Weather-Corrected Performance Ratio (WCPR)

According to the NREL [8], the WCPR is a temperature-corrected performance index that helps reduce fluctuations by correcting for the temperature loss caused by solar PV cells in the solar array.

$$WPCR = PR \frac{1}{1+\delta\left(T_{cell,avg} - T_{cell}\right)} \tag{1.2}$$

where the solar cell's temperature coefficient is denoted by δ; the yearly average temperature of a solar cell is represented by $T_{cell, avg}$; and the solar cell's temperature is indicated by T_{cell}.

Equation 1.1 is an estimated equation for computing the temperature of the solar PV cell,

$$T_{cell} = T_{amb} + \frac{G}{800}\left(T_{NOCT} - 20\right) \tag{1.3}$$

where T_{amb} represents the ambient temperature, and T_{NOCT} represents the Normal Operating solar PV cell temperature.

1.6.1.3 Linear Regression (LR) and Weather-Corrected Linear Regression (WCLR)

When sun irradiation is minimal, the PR index has difficulties in classifying the fault condition. Given the enormous lack of precision introduced by the PR index due to the unexpectedness of solar resource at low irradiation, LR delivers a more accurate representation of the vast variety of input factors and minimizes environmental resource variability. LR provides a better representation. Additionally, WCLR stabilizes the linear pattern by accounting for a solar PV cell's temperature loss. A solar PV power plant's coefficient of determination R^2 is equal to zero when a failure occurs in the system. As a matter of fact, the computed value of R^2 in normal operation shows that there is a strong correlation between the amount of radiation and the amount of power.

$$P_{AC,WC} = \frac{P_{AC}}{1 + \delta\left(T_{cell,avg} - T_{cell}\right)} \tag{1.4}$$

where $P_{AC,WC}$ is the solar PV power plant's output power that has been corrected for temperature loss.

1.7 ASSESSMENT OF SOLAR POWER PLANT RESOURCES

1.7.1 Clearness Index (CI)

A numerical evaluation of the clarity level of the atmosphere can be accomplished by using a clearness index, which fluctuates depending on the weather circumstances, such as cloudiness or clear skies [28]. This indicator shows the fraction of solar energy that makes it to the Earth's surface, with zero representing no solar energy and 1 representing 100%. When evaluating a clearness index, it is preferable to make a precise numerical assessment rather than an uncertain distinction like a sunny day vs. an overcast day [30].

$$sd = 23.45 \sin\left[\left(\frac{360}{365}\right).\left(n - 81\right)\right] \tag{1.5}$$

where n is the number of days in the year and sd is the solar declination that varies from 1 to 365. Equation 1.6 represents the day-to-day variations in extraterrestrial solar insolation.

$$I_0 = SC\left[1+0.034\cos\left(\frac{360n}{365}\right)\right] \tag{1.6}$$

where SC represents the solar constant (1367 W/m^2). Equation 1.7 expresses the sunrise hour angle (in radians)

$$H_{SR} = \cos^{-1}\left(-\tan L.\tan\delta\right) \tag{1.7}$$

where L denotes the solar latitude angle. To find out the daily extraterrestrial sunlight insolation (H_0) on a horizontal surface, you need to add the product of the normal radiation and the sine of the solar altitude angle, and then multiply that number by the number of hours in a day. This equation is given in Equation 1.8.

$$H_0 = \frac{24}{\Pi} I_0\left(\cos L.\cos\delta.\sin H_{SR} + H_{SR}\sin L.\sin\delta\right) \tag{1.8}$$

The first stage in deconstructing total horizontal insolation into its diffuse and beam components is to define a clearness index (K_t), which is the ratio of average horizontal insolation at a place to extraterrestrial insolation on a horizontal surface above and just outside the atmosphere.

$$K_t = \frac{H}{H_0} \tag{1.9}$$

A high value of K_t indicates clearer skies with almost all of the radiation being direct beam, whereas a low K_t is associated with cloudy skies and has almost all of the radiation being diffused.

1.7.2 Proposed Variable Index (PVI)

A solar PV power plant's power generation is lowered significantly as a result of the fluctuation of cloud coverage. In response to the variation in natural resources, the output power of renewable energy sources has a wide range of fluctuations. When the variability value is determined precisely, the performance of solar PV power plants may be assessed objectively, bearing in mind the reduction in energy generation owing to variability [31]

$$PVI = \frac{1}{n-1}\sum_{k=2}^{n}\left|\left(\frac{P_{AC,k}}{P_{DC,k}}\right)^2 - \left(\frac{P_{AC,k-1}}{P_{DC,k-1}}\right)^2\right| \tag{1.10}$$

where $P_{AC,k}$ and $P_{DC,k}$ correspond to the k state of AC & DC power output from the solar PV plant, and $P_{AC,k-1}$ and $P_{DC,k-1}$ correspond to the $k-1$ state of the solar PV plant's AC & DC power output.

1.8 TECHNIQUE FOR DETECTING OUTLIERS IN SOLAR PV POWER PLANTS

1.8.1 Estimated Error Matrix (EEM)

The difference between the measured and predicted values of the regression analysis is represented by the estimated error. The EEM generates a big value in cases when a fault occurs, because the linear regression equation deviates from the actual measurement by a large amount. Furthermore, the EEM can provide a greater score for a defect on a clearer day and a lower value for a fault with decreased output power due to fluctuation [31].

$$EEM = \begin{bmatrix} \hat{Y}_1 - Y_1 \\ \vdots \\ \vdots \\ \hat{Y}_n - Y_n \end{bmatrix} ./ \begin{bmatrix} Y_1^{-1} \\ \vdots \\ \vdots \\ Y_n^{-1} \end{bmatrix} \tag{1.11}$$

where $\hat{Y}_i$ represents the estimated value of the solar PV plant's output power and Y_i represents the observed value of the solar PV plant's output.

The EEM signifies all the values in the data sets in $n \times 1$ by using a matrix representation. Here, n represents the solar power plant data collected period. When a solar panel maintenance engineer extracts the output power of a solar plant over a one-year period, a 365×1 matrix is formed.

1.8.2 Estimated Square Error Index (ESEI)

An $n \times n$ matrix is formed by multiplying the estimated error matrix by a factor of two. This is known as an ESE matrix. A diagonal matrix is used to calculate the distance between each output, which can be stated as an ESEI. If a fault occurs on a photovoltaic system, because of the high error value and the linear algebra technique utilized, the magnitude of ESEI is bigger than other values [31]

$$ESEM = EEM.EEM' \tag{1.12}$$

$$ESEI = diag\left(ESEM\right) \tag{1.13}$$

1.8.3 Support Vector Machine (SVM)

The SVM is a mostly preferred approach for classifying multi-labeled data sets by analyzing point data. The algorithm uses a linear model to find the largest margin between training and test data samples. Using the Support Vector Machine (BCSVM), solar PV output power data can be classified as "fault" or "non-fault" [31]

$$\min_{w,b,\alpha} \frac{1}{2}\left|w^2\right| = \sum_{i=1}^{N} \alpha_i \left[y_i \left(\overrightarrow{w}.\overrightarrow{x_i} + b \right) - 1 \right] \tag{1.14}$$

$$\max_{a} L(\alpha_i) = \sum_{i=1}^{N} \alpha_i - \frac{1}{2} \sum_{i,j=1}^{N} \alpha_i \alpha_j y_i y_j x_i^T x_j \tag{1.15}$$

where w represents vector, x_x and y_x corresponds to the data set variables, and α stands for Lagrange multiplier.

1.8.4 Kernel Function

A high-dynamic, implicit feature space can be created using the kernel function, although no data should ever be stored in smaller spaces. The problem of misclassification in lower-dimensional data sets is effectively resolved. A feature vector is created by transferring the raw data in lower space to a feature vector [31]

$$f(x_i, x_j) = \phi(x_i).\phi(x_j) \tag{1.16}$$

where the kernel function is represented by ϕ, and x_i & x_j correspond to the input variables of input space. Below is a list of the kernel functions used in this technique.

Polynomial function:

$$f(x_i, x_j) = (x_i . x_j + 1)^p \tag{1.17}$$

Gaussian function:

$$f(x_i, x_j) = \exp\left(-\frac{(x_i - x_j)^2}{2\sigma^2} \right) \tag{1.18}$$

Sigmoid function:

$$f(x_i, x_j) = \tan\text{h}\left(\alpha x_i^T y_i + c \right) \tag{1.19}$$

where p, σ, α, and c correspond to the degree of the polynomial, standard deviation, slope, and intercept constant, respectively.

1.9 CONCLUSION

In this chapter, the necessity of renewable energy sources for the production of electricity compared to other fossil fuels has been discussed. Solar is one of the prominent technologies for the upcoming years. Even though it is developing, the solar power plant faces some challenges in day-to-day life. In this chapter, the AI approaches for predicting the faults and identifying the reason for the reduction of output power have been discussed, as well as the types of fault states identified by using AI techniques. The performance of solar plants and their aging are detected by the PR index, but due to variation in temperature, its efficiency gets reduced. Therefore, it is prevailed over by WCPR value and stabilizing the output power and by compensating for the temperature variation of PV modules. By comparing the output variables of solar plant with input variables through a kernel function used to identify the exact fault can improve the output power. This type of online fault diagnosis is used to identify the fault in a solar power plant with less time, and this can be further improved by using thermal imaging techniques capable of getting maximum output power generated from a solar PV plant.

REFERENCES

1. Milligan, M., Frew, B., Kirby, B., Schuerger, M., Clark, K., Lew, D., ... & Tsuchida, B. (2015). Alternatives no more: Wind and solar power are mainstays of a clean, reliable, affordable grid. *IEEE Power and Energy Magazine*, *13*(6), 78–87.
2. IEC 61724-1:2017. Photovoltaic System Performance—Part 1: Monitoring. Geneva, Switzerland: IEC, 2017.
3. Basson, H. A., & Pretorius, J. C. (2016, May). Risk mitigation of performance ratio guarantees in commercial photovoltaic systems. In *International Conference Renewable Energies and Power Quality*, Madrid, Spain.
4. Urrejola, E., Antonanzas, J., Ayala, P., Salgado, M., Ramírez-Sagner, G., Cortés, C., ... & Escobar, R. (2016). Effect of soiling and sunlight exposure on the performance ratio of photovoltaic technologies in Santiago, Chile. *Energy Conversion and Management*, *114*, 338–347.
5. Bizzarri, F., Brambilla, A., Caretta, L., & Guardiani, C. (2015). Monitoring performance and efficiency of photovoltaic parks. *Renewable Energy*, *78*, 314–321.
6. Ventura, C., & Tina, G. M. (2016). Utility scale photovoltaic plant indices and models for on-line monitoring and fault detection purposes. *Electric Power Systems Research*, *136*, 43–56.
7. Chine, W., Mellit, A., Pavan, A. M., & Kalogirou, S. A. (2014). Fault detection method for grid-connected photovoltaic plants. *Renewable Energy*, *66*, 99–110.
8. Dierauf, T., Growitz, A., Kurtz, S., Cruz, J. L. B., Riley, E., & Hansen, C. (2013). *Weather-corrected performance ratio* (No. NREL/TP-5200-57991). National Renewable Energy Lab. (NREL), Golden, CO (United States).
9. Mellit, A., & Kalogirou, S. A. (2018). A survey on the application of artificial intelligence techniques for photovoltaic systems. In Soteris A. Kalogirou (ed.) *McEvoy's Handbook of Photovoltaics* (pp. 735–761). London: Academic Press.
10. Mahajan, V. (2014). *PV Module and system fault analysis* (Doctoral dissertation, Murdoch University).
11. Cunningham, D. W., & Solar, B. P. (2011). Analysis of hot spots in crystalline silicon modules and their impact on roof structures. In *Photovoltaic Module Reliability Workshop 2011 (PVMRW)* (p. 642). Denver, CO, USA: National Renewable Energy Laboratory.
12. Sinha, A., Sastry, O. S., & Gupta, R. (2016). Detection and characterisation of delamination in PV modules by active infrared thermography. *Nondestructive Testing and Evaluation*, *31*(1), 1–16.

13. Abdelhamid, M., Singh, R., & Omar, M. (2013). Review of microcrack detection techniques for silicon solar cells. *IEEE Journal of Photovoltaics*, *4*(1), 514–524.
14. Patel, H., & Agarwal, V. (2008). MATLAB-based modeling to study the effects of partial shading on PV array characteristics. *IEEE Transactions on Energy Conversion*, *23*(1), 302–310.
15. Zhao, Y., Lehman, B., de Palma, J. F., Mosesian, J., & Lyons, R. (2011, September). Challenges to overcurrent protection devices under line-line faults in solar photovoltaic arrays. In *2011 IEEE Energy Conversion Congress and Exposition* (pp. 20–27), IEEE.
16. Flicker, J., & Johnson, J. (2016). Photovoltaic ground fault detection recommendations for array safety and operation. *Solar Energy*, *140*, 34–50.
17. Takashima, T., Yamaguchi, J., & Ishida, M. (2008). Disconnection detection using earth capacitance measurement in photovoltaic module string. *Progress in Photovoltaics: Research and Applications*, *16*(8), 669–677.
18. Patel, H., & Agarwal, V. (2008). MATLAB-based modeling to study the effects of partial shading on PV array characteristics. *IEEE Transactions on Energy Conversion*, *23*(1), 302–310.
19. Spataru, S., Sera, D., Kerekes, T., & Teodorescu, R. (2012, June). Detection of increased series losses in PV arrays using Fuzzy Inference Systems. In *2012 38th IEEE Photovoltaic Specialists Conference* (pp. 000464–000469). IEEE.
20. Yuventi, J. (2013). DC electric arc-flash hazard-risk evaluations for photovoltaic systems. *IEEE Transactions on Power Delivery*, *29*(1), 161–167.
21. Chang, M., Chen, C., Hsueh, C. H., Hsieh, W. J., Yen, E., Ho, K. L., ... & Chen, H. (2015, June). The reliability investigation of PV junction box based on 1GW worldwide field database. In *2015 IEEE 42nd Photovoltaic Specialist Conference (PVSC)* (pp. 1–4). IEEE.
22. Alajmi, M., & Abdel-Qader, I. (2016, May). Fault detection and localization in solar photovoltaic arrays using the current-voltage sensing framework. In *2016 IEEE International Conference on Electro Information Technology (EIT)* (pp. 307–312). IEEE.
23. Stellbogen, D. (1993, May). Use of PV circuit simulation for fault detection in PV array fields. In *Conference Record of the Twenty Third IEEE Photovoltaic Specialists Conference-1993 (Cat. No. 93CH3283-9)* (pp. 1302–1307). IEEE.
24. Spataru, S., Sera, D., Kerekes, T., & Teodorescu, R. (2015). Diagnostic method for photovoltaic systems based on light I–V measurements. *Solar Energy*, *119*, 29–44.
25. Chouay, Y., & Ouassaid, M. (2017, November). An intelligent method for fault diagnosis in photovoltaic systems. In *2017 International Conference on Electrical and Information Technologies (ICEIT)* (pp. 1–5). IEEE.
26. Coleman, A., & Zalewski, J. (2011, September). Intelligent fault detection and diagnostics in solar plants. In *Proceedings of the 6th IEEE International Conference on Intelligent Data Acquisition and Advanced Computing Systems* (Vol. 2, pp. 948–953). IEEE.
27. Takashima, T., Yamaguchi, J., Otani, K., Kato, K., & Ishida, M. (2006, May). Experimental studies of failure detection methods in PV module strings. In *2006 IEEE 4th World Conference on Photovoltaic Energy Conference* (Vol. 2, pp. 2227–2230). IEEE.
28. Mellit, A., Tina, G. M., & Kalogirou, S. A. (2018). Fault detection and diagnosis methods for photovoltaic systems: A review. *Renewable and Sustainable Energy Reviews*, *91*, 1–17.

29. Chouder, A., & Silvestre, S. (2010). Automatic supervision and fault detection of PV systems based on power losses analysis. *Energy Conversion and Management*, *51*(10), 1929–1937.
30. Masters, G. M. (2013). *Renewable and efficient electric power systems*. New York: John Wiley & Sons.
31. Shin, J. H., & Kim, J. O. (2020). On-line diagnosis and fault state classification method of photovoltaic plant. *Energies*, *13*(17), 4584.

2 Fault Diagnosis Techniques for Solar Plant Based on Unsupervised Sample Clustering Probabilistic Neural Network Model

K. Umamaheswari and N. Yogambal Jayalashmi
Dr. Mahalingam College of Engineering and Technology,
Pollachi, India

CONTENTS

LEARNING OUTCOME

i. To study the various types of faults occurring in PV systems.
ii. To develop a novel fault diagnosis strategy to enhance the performance of the PV systems.
iii. To optimize the performance of PV systems using artificial intelligence techniques.

DOI: 10.1201/9781003202288-2

2.1 INTRODUCTION: BACKGROUND AND DRIVING FORCES

The solar energy is the one of the crucial sources of clean energy. In photovoltaic power generation the intensity of light energy is converted into electricity by PV modules. The basic principle of operation is that when the photons of light impact the photovoltaic cells, the electrons are released and circulated. The generated power is in the form of direct current (DC), and the inverter system is employed to convert it into alternating current (AC) for industrial and residential purposes.

The solar panel is a device that possesses an array of numerous solar cells that capture the photons from sunlight and convert them to electricity. They carry an anti-reflection coating to reduce the loss of photons. The main challenge of the photovoltaic power generation is its vulnerability to harsh environmental conditions, and the detection and mitigation of faults are an inevitable field of research.

2.1.1 Existing Works of Literature

The efficiency of the photovoltaic power generation can be improved by early fault identification and mitigation steps (Qureshi et al. 2020); an independent component analysis (ICA) is employed for early prognosis of fault. The evolution of photovoltaic components requires advanced fault identification and mitigation techniques (Triki-Lahiani et al. 2018). To reduce the fault scenarios in PV systems, numerous standards are employed, such as International Electrotechnical Commission (IEC), National Eical Code (NEC), and Underwriters Laboratories (UL) (Mellit et al. 2021), so IoT-based fault diagnosis techniques are implemented along with deep learning algorithms for early identification. The fuzzy logic–based multilayered photovoltaic detection is implemented based on LabVIEW simulation techniques (Dhimish et al. 2017), and for grid-connected fault detection, a statistical t-test based LabVIEW system software is employed to identify the faulty MMP units (Dhimish and Holmes 2016). To locate the fault, a multilevel wavelet decomposition strategy is employed (Kim 2016). The photovoltaic fault conditions such as PV module fault, faulty PV string, bypass diode fault, and MMPT fault and DC/AC inverter units are identified (Dhimish et al. 2017). A clustering strategy is formulated to identify the art and ground fault conditions based on a robust minimum covariance estimator (Braun et al. 2012). Incipient changes occurring in PV modules are estimated by an exponentially weighted moving average (EWMA) model, which is employed to identify the type of fault occurring in grid-connected PV modules (Garoudja et al. 2017). An artificial neural network strategy based on a radial basis functional neural network strategy is employed to identify the partial shading condition with reduced number of input parameters (Hussain et al. 2020). Various neural network techniques have been implemented for early fault detection (Li et al. 2020). On utilizing the current voltage characteristics of PV arrays, the line-to-line fault of PV systems on DC side is identified based on probability strategy with better performance; here several learning strategies are combined to perform accurate fault detection (Eskandari et al. 2020), whereas the ensemble of numerous learning strategy increases computational complexity. The fault detection methods is employed not only for fault identification, in order to improve the efficiency and effectiveness; it should also be employed to treat some undesirable faults, as proposed by Bendary (2021), in whose study a fuzzy logic–based adaptive neuro fuzzy model is developed to identify, mitigate, and clear the fault if any occur. During fault scenario, a new data stream is transmitted from PV arrays to the diagnosis system, where a density-based spatial clustering of applications with noise (DBSCAN) algorithm is developed to perform an online fault diagnosis (Cai et al. 2020).A neural network–based fault detection strategy is developed to identify the short-circuit fault, the healthy sections, and the disconnection of strings in PV array (Khelil et al. 2020). The feature selection is an important strategy to be followed to improve the performance of PV fault diagnosis models; it has advantage of reducing the input data dimensions so the efficiency of the model performance can be improved (Hajji et al. 2020). To select the essential features, a principal component technique is employed over supervised machine learning algorithms for early fault detection in PV systems. For large-scale grid-connected PV systems, the fault detection and mitigation are a challenging task, so a decomposed open-close alternating sequence technique is developed to correlate the random spikes with the possibility of arch occurrences. Machine learning algorithms are employed to identify the PV fault such as short-circuit, open-circuit, partial shadowing, and degradation faults (Lazzaretti et al. 2020). To improve the performance of the PV fault detection techniques, the nonlinear behavior of the system should be handled effectively, so a sequential probabilistic neural network is developed to handle the nonlinearity of PV systems; here ten different faults are identified (Zhu et al. 2020). The output deviation of PV array is identified by PNN (Zhao et al. 2020). A reinforcement learning strategy is presented for fault detection and diagnosis based on current voltage characteristics of the photovoltaic array (Zhang et al. 2020). The PV array fault is identified based on feed forward neural networks with cyber physical system approach (Rao et al. 2019). The common faults of PV systems such as line-to-line fault, open-circuit fault, partial shading fault, and bypass diode fault are identified by a K-NN strategy with current voltage analyzer module (Madeti et al. 2018).

On analyzing the numerous strategies in various works of literature, it is identified that the high-nonlinear operating condition of PV modules increases the challenging of fault identification. Even though numerous machine learning deep learning models are developed to address the problem of fault identification, it is an open field of research that requires a highly efficient model with less complexity to perform early and effective fault diagnosis. Based on the above survey, the Probabilistic Neural Network (PNN) model finds high advantage in the field of classification, so in this chapter a PNN-based PV-fault diagnosis technique is developed and discussed in the following sections.

2.2 THEORY OF PHOTOVOLTAIC FAULTS

The faults of the photovoltaic system are categorized into two main types, the temporary and the permanent, as shown in Figure 2.1. The temporary faults will occur for a certain period and can be cleared by design. Whereas the permanent fault is severe and persistent, special equipment is necessary to identify the faults. Each time partial shading occurs and is left untreated, the entire power system collapses. Numerous numbers of PV modules are connected in parallel to form a PV module, and the PV string is constructed by connecting numerous PV modules in series. The PV strings are connected in parallel to form a PV array. The PV cells are connected via a diode in

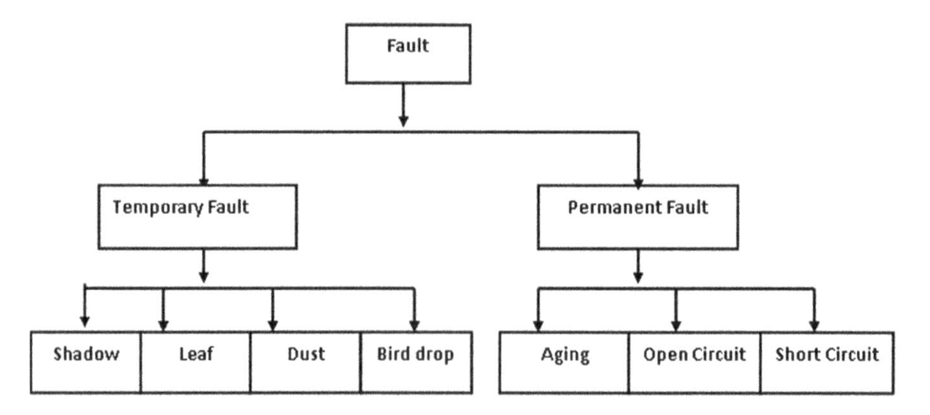

FIGURE 2.1 Types of fault occurrence in photovoltaic system.

parallel, so that the hotspot phenomenon of the PV module can be avoided. To improve the PV module output, numerous PV modules are connected in series by an n-cell.

As the fault occurs, the temperature difference arises between the healthy and unhealthy sections, which is exacerbated if partial shading arises at the terminal. The cell cracks will take place in case of excessive thermal risk, and considerable damage will occur if reverse bias voltage exceeds the breakdown cell voltage.

2.2.1 Photovoltaic Minor Fault Condition

The PV power panel can operate during minor fault conditions, and the faulty section will be powered by the healthy section. Consider a PV array, which consists of PV modules of 'n' rows and 'm' columns. The voltage, current, and the power profile of the PV array are represented as 'v_arry', 'I_arry', and 'P_arry'. The current and voltage profiles for healthy PV module strings are represented by 'v_h' and 'v_f', respectively, and the current is represented as 'I_h' and 'I_f'. The healthy and faulty string temperature is represented as T_h and T_f, respectively. The faulty cell in the array will become a resistive load R_f, and the other healthy cells will supply the entire load including the faulty string. The faulty string electrical characteristics are presented as follows:

$$V_h - I_f R_f = V_{arry} \tag{2.1}$$

$$I_f = \frac{V_h}{R_f + R_{load}} \tag{2.2}$$

$$\Delta V = V_h - V_f \tag{2.3}$$

$$\Delta I = I_h - I_{fh} \tag{2.4}$$

$$I_f^2 R_f < I_f \left(x - y\right) \frac{V_f}{x.y} \tag{2.5}$$

where x and y specify the number of healthy and faulty strings; here the generated power is supplied both to the load and the faulty cells.

2.2.2 Photovoltaic Medium Fault Condition

The working condition of the PV module strings is affected by the medium fault. During medium fault, each of the PV module possesses different operating current, which results in multistage characteristics. The faulty sections are short-circuited in the high output current operating region.

2.2.3 Photovoltaic Heavy Fault Condition

During heavy fault, the bypass diode is short-circuited and the healthy section is open-circuited. The bypass diode is connected in anti-parallel, the faulty module is shorted, and the healthy sections are open-circuited. Thus, the output power is less than the actual power delivered to the load.

2.3 MATHEMATICAL MODELING OF THE PV SYSTEMS

In this chapter, the PV fault detection is made based on the real-time data simulated in MATLAB® environment. Modeling of the healthy PV system is presented as follows:

$$I = I_L - I_o\left(\exp\left(\frac{\eta V}{T_m}\right) - 1\right) \tag{2.6}$$

$$\mu = \frac{q}{N_s K \varepsilon} \tag{2.7}$$

$$I_L = \frac{G}{G_{ref}}\left[I_{lref} + k_i\left(T_m - T_{ref}\right)\right] \tag{2.8}$$

$$I_o = I_{oref}\left(\frac{T_m}{T_{ref}}\right)^3 \exp\left(\frac{q}{N_s}\cdot\frac{E_{BG}}{\varepsilon K}\right)\left(\frac{1}{T_{ref}} - \frac{1}{T_m}\right) \tag{2.9}$$

where the output current is represented by I, the photon current is represented by I_L, q specifies the charge, the diode characteristic fault is ε, the Boltzmann constant is K, the irradiance is G, the reference irradiance is G_{ref}, Ns specifies the number of series-connected cells, the reference values of current are represented by I_{Lref}, reference values of saturated current are given as I_{oref}, saturated current is I_o, the current temperature coefficient is k_i, the reference temperature is T_{ref}, and the module temperature is T_m.

$$I_{sc\_ref} - I_{mppt\_ref} = \frac{I_{sc\_ref}}{\exp\left(\frac{\mu.V_{oc\_ref}}{T_{ref}}\right) - 1}\exp\left[\left(\frac{\mu.V_{mpp\_ref}}{T_{ref}}\right) - 1\right] \tag{2.10}$$

where I_{mppt_ref} and V_{mppt_ref} are, respectively, current and voltage of maximum power point tracking, I_{sc_ref} is the short circuit, and V_{oc_ref} refers to the open-circuit voltage reference condition.

2.4 MATERIALS AND METHOD

In this proposed fault diagnosis strategy, the fault classification is performed using the probabilistic neural network technique. To perform effective classification, the feature selection is performed by the proposed hybrid SSA strategy of combining the characteristics of Multiverse Optimization Algorithm (MOA) and Social Spider Optimization Algorithm (SSA). The experimental modeling is made using the real-time data set simulated in MATLAB® software. The proposed methodology framework is presented in Figure 2.2.

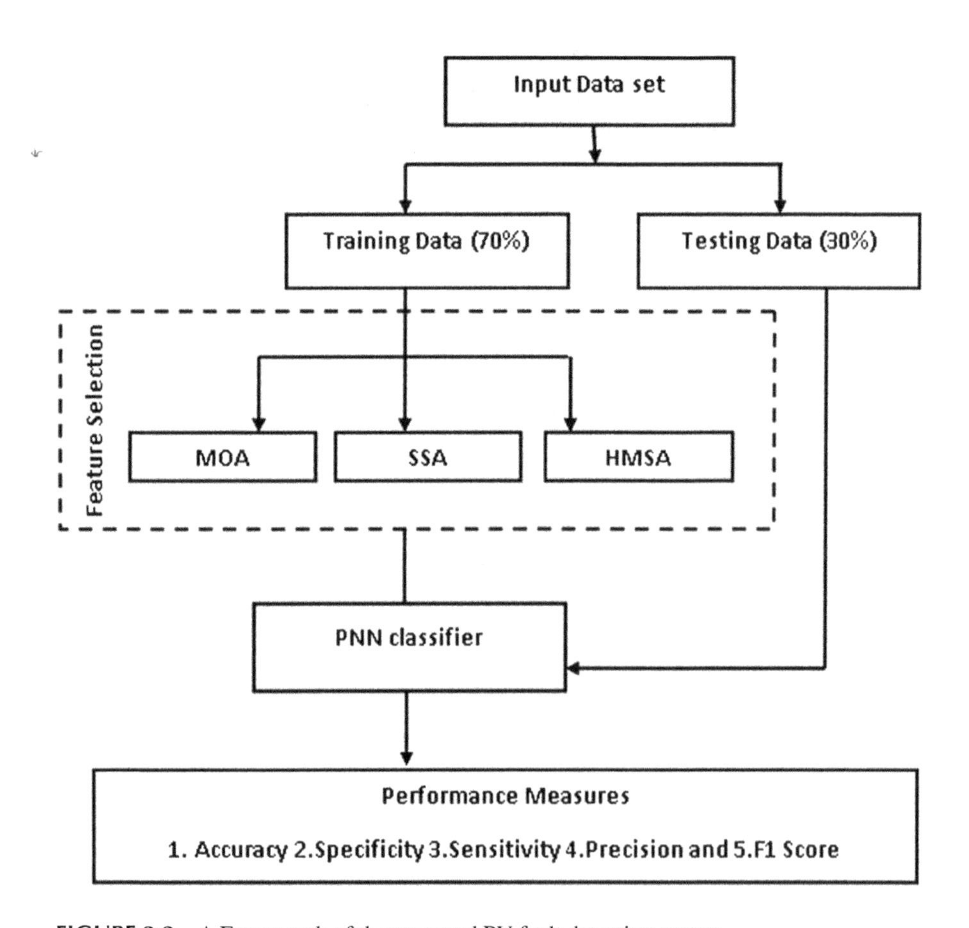

FIGURE 2.2 A Framework of the proposed PV fault detection system.

2.4.1 Probabilistic Neural Network

The probabilistic neural network (PNN) is a feed-forward neural network, introduced by Donald Seecht in 1988, and predominantly employed for classification and pattern recognition applications. It is a four-layered architecture whose learning mechanism is based on Bayes decision strategy and probability density functions. Here, the Gaussian kernel function is employed to identify the necessary features of each class during the learning process. The estimated densities are adopted over the Bayes decision function in order to complete the classification process. The smooth and continuous operation of the probability density function is the advantage of PNN.

The architecture of PNN is presented in Figure 2.3; the input layer is the very first layer of the architecture provided with an input vector corresponding to the number of features of the problem. In the hidden layer the pattern neuron computes the centroid of the feature vector by estimating the Euclidean distance from the training sample and the input feature vector. In the summation layer, the sum of output of all hidden layers is estimated and the output of highest value is passed into output layer through a decision function.

The PNN is the neural implementation of the parsed windows, which is a non-parametric recognition framework that synthesizes the computation of the probability density function by super position of replicas of a kernel function $g(.)$,

$$f(x) \cong f_n(x) = \frac{1}{n\lambda}\sum_{i=1}^{n} g\left(\frac{x - x(l)}{\lambda}\right) \tag{2.11}$$

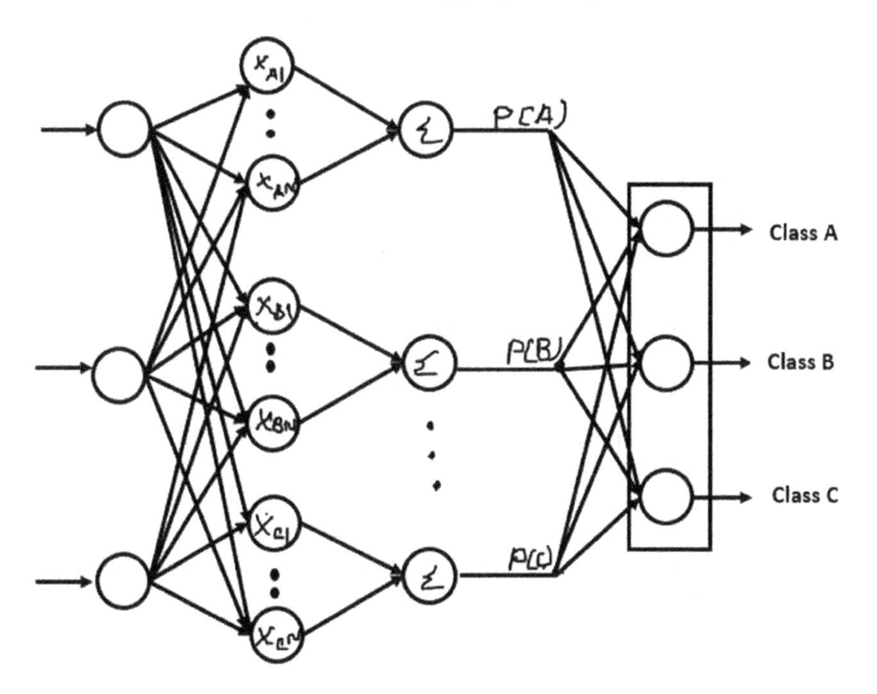

FIGURE 2.3 Architecture of probabilistic neural network model.

where x is the replica argument for a position in search space, $x(l)$ forms the training sample, and λ is a function of n such that

$$\lim_{n\to\infty} \lambda = 0 \text{ and} \tag{2.12}$$

This calculation is consistent in quadratic mean,

$$\lim_{n\to\infty} E\left|f(x) - f_n(x)\right|^2 = 0 \tag{2.13}$$

where the function $f(x)$ is continuous. Consistency indicates that this is a modest approximation method, such that when the number of training samples grows, the estimate tends smoothly to the actual density. The optimal classification strategy employed to achieve the Bayesian decision criterion is the optimal classification procedure if the probability density functions of the classes to be discriminated are known.

The following Bayesian technique can be applied,

$$c = \arg\max_k \left\{\gamma^k f^k(x)\right\} \tag{2.14}$$

where c is the class label output by the classifier, x is a vector in the input space, $f^k(x)$ is the density of the k-th class, and the coefficient γ^k contains the prior probability and the risk coefficients, if any associated with that class.

The training of PNN is made by a quick strategy that involves only localized parameter modifications; this training procedure is not same as other conventional training algorithms such as PNN. The PNN learning strategy comprises of introducing a pattern sample for each original learning sequence varying its input weight so that they coincide with the pattern itself. The following activation function is considered to introduce the preferred kernel function, positioned on the certain pattern. There is no tunable parameter for the summation layer, and the subsequent frequency of training pattern appears at the output layer. Thus, the training is a one-step process, which does not need any iterations, which makes it a quick learning process.

2.4.2 Multiverse Optimization Algorithm (MOA)

The presence of more than the one universe in which we are living is the multiverse concept, on the basis of which Mirjalili developed the multiverse optimization algorithm in 2016. All the universes are interconnected with each other through black holes, white holes, and wormholes. The solution is represented by each element of the universe; if the probability of white holes is higher and black holes is lower, the inflation rate of the universe increases. The path from a white hole to a black hole tends to transfer the objects in the universe. The solution is obtained when the object searches and finds the best universe. If the inflation rate of a black hole is lesser than that of a white hole, the object is transferred from white holes to black holes. The objective of the algorithm is to improve the average inflation rate over the iterative process. Based on the inflation rate, the while hole is arranged to each universe through an iterative process.

The population matrix is defined by the matrix,

$$X_{pop} = \begin{bmatrix} x_1 & . & . & x_1^d \\ x_2 & . & . & x_2^d \\ . & . & . & . \\ x_n & . & . & x_n^d \end{bmatrix} \quad (2.15)$$

The X_{pop} is the universe population with n universes of d dimensions. The decision variable is presented as

$$x_i^j = \left\{ lowerbd_j + rand(\ )\left(\left(upperbd_j - lowerbd_j\right) + 1\right) \right. \quad (2.16)$$

The $lowerbd_j$ and $upperbd_j$ represent the bound limit of the decision variables, and the *rand*() attains the value 0 to 1. Based on the inflation rate, the position of a black hole varies over the iterations as presented in the following equation:

$$x_i^j = \begin{cases} x_{np}^j\ rand1(\ ) < Norm(U_i) \\ x_i^j\ rand1(\ ) \geq Norm(U_i) \end{cases} \quad (2.17)$$

where x_i^j is the j-th decision variable of the i-th universe, *Ui* is the i-th universe, $Norm(U_i)$ is the normalized inflation rate of i-th universe, *rand1( )* is a randomly generated number in the [0,1] range, and x_{np}^j is the j-th variable of n-th universe. The solution of the wormhole is given by the following expressions:

$$x_i^j = \begin{cases} \begin{cases} x_{fit\_j} + \eta \times \left(\left(upperbd_j - low\_bd_j\right) \times r1 + lowerbd_j\right) r2 < 0.5 \\ x_{fit\_j} - \eta \times \left(\left(upperbd_j - low\_bd_j\right) \times r1 + lowerbd_j\right) r2 \geq 0.5 \end{cases} & r3 < P_{wh} \\ & r3 \geq P_{wh} \end{cases} \quad (2.18)$$

where x_{fit_j} is the j-th fittest universe variable, the minimum bound limit and maximum bound limit of j-th parameter are given by $lowerbd_j$ and $upperbd_j$, respectively, with x_i^j indicating the j-th parameter of the i-th universe, $r1$, $r2$, and $r3$ being random numbers from 0 to 1, η being the rate of distance traveled, and P_{wh} indicating the probability of the presence of wormholes. The adaptive variation of the parameters ρ and P_{wh} is given by

$$\rho = 1 - \frac{\left(current\_iteration\right)^{1/ea}}{\left(\max\_iteration\right)^{1/ea}} \quad (2.19)$$

$$P_{wh} = C_{\min} + current\_iteration \times \left(\frac{C_{\max} - C_{\min}}{\max\_iteration} \right) \quad (2.20)$$

where *ea* represents the exploitation accuracy during the iterative process, and *Cmin* and *Cmax* indicate the predefined constant values that help in the exploitation phase during iterations. The higher the value of *ea*, the more accurate and faster is the exploitation mechanism.

2.4.3 Social Spider Optimization Algorithm (SSA)

One of the swarm intelligence–based algorithms is the social spider optimization algorithm that is based on the social intelligence of spiders that live in a colony (Cuevas et al. 2013). Spiders in the colony communicate among themselves through vibrations that occur in the string that they form. In this algorithm, the search space is the "spider web" and the search agents are the "spiders" that move through the colony. The search agents get divided into male agents and female agents, and each spider's weight determines the objective function value. The candidate solution in the search space is determined based on the gender of the spider. The number of female spider forms is 60% to 90% of the total population. The size of the spider that sends the information and the distance at which it presents determine the vibration that a spider receives.

The vibration coefficient is given by

$$Vbr\_s_m = w_n e^{d_{m,n}^2} \tag{2.21}$$

In the above equation, w_n specifies the weight of the n-th spider and $d_{m,n}$ indicates the Euclidean distance between m-th and n-th spiders. Spiders in the web come across three vibrations – the nearest spider possesses higher-fitness V_{br_hm}, the best spider is denoted as V_{br_bm}, and the nearest female that is applicable for males is specified as V_{br_fm}. The female update position is given by

$$X_m(t+1) = X_m(t) + \begin{pmatrix} \alpha.Vbr\_h_m.\left(F_c - X_m(t)\right) + \beta.Vbr\_b_m.\left(F_b - X_m(t)\right) \\ +\delta.\left(\lambda - \frac{1}{2}\right) \end{pmatrix} \tag{2.22}$$

In equation (2.22), α, β, δ, and λ refer to random numbers from 0 to 1 and are control parameters, *Fc* specifies the best closest neighbor, and the fittest spider in the swarm is *Fb*. When the female spiders want to move away from the source, it is given by

$$X_m(t+1) = X_m(t) - \begin{pmatrix} \alpha.Vbr\_h_m.\left(F_c - X_m(t)\right) + \beta.Vbr\_b_m.\left(F_b - X_m(t)\right) \\ +\delta.\left(\lambda - \frac{1}{2}\right) \end{pmatrix} \tag{2.23}$$

The update for male spiders is different from that of the female spiders; the male spider's position is updated based on whether they are dominant or nondominant spiders. The position update for nondominant *Xm_ND* and dominant *Xm_D* spider is respectively given with equations (2.24) and (2.25).

$$X_{m\_ND}(t+1) = X_m(t) + \left(\alpha.Vbr\_f_m.\left(F_f - X_m(t)\right) + \delta.\left(\lambda - \frac{1}{2}\right)\right) \tag{2.24}$$

$$X_{m\_D}(t+1) = X_m(t) + \left(\alpha . \left(\frac{\sum_{n=1}^{N_m} \omega_n^k . W_{N_m} + n}{\sum_{n=1}^{N_m} W_{N_m}} \right) - X_m(t) \right) \quad (2.25)$$

where *Ff* specifies the closest female with respect to the n-th make and β, δ, and λ are random numbers from 0 to 1. The nondominant spiders are only capable of moving toward the females, and the final step in SSA technique is the mating operation that modifies the search agents. Within the mating radius the nondominant males mate with the females, which is accompanied by a roulette-wheel selection, as there might be more than one male and one female within the radius of mating. The new offspring is generated by gene recombination of fittest parents, and at each iteration the least-fit spiders are removed from the population.

2.4.4 Hybrid Multiverse Social Spider Optimization Algorithm (HMSA)

The social spider optimization algorithm is performed effective on feature selection, but over the iterations the convergence speed decreases, and in some cases premature convergence has been reported. To handle this limitation, a multiverse optimization algorithm is hybridized with it. The basic strategy behind hybridization is that in case the SSA struggles to find the optimal solutions for the considered problem, the MOA is utilized, and the initial population for the current iteration of MOA is the fittest population obtained from the previous iteration of an SSA optimizer. The effectiveness of the given algorithm is evaluated by means of its efficiency rate compared to other conventional optimization algorithms. The input data set is segregated into 10 sets, and for training each data set, the efficiency rate of SSA is compared to other optimization algorithms such as Genetic Algorithm (GA), Particle Swarm Optimization (PSO), Biogeography-Based Optimization (BBO), The Flower Pollination Algorithm (FPA), Grey Wolf Optimization (GWO), Bat Optimization Algorithm (BAT), Fire Fly Optimization Algorithm (FA), Cuckoo Search Optimization (CS), Moth Flame Optimization (MFO), Teaching Learning Based Optimization Algorithm (TLBO). To avoid biased output, the algorithm is made to run 100 times and the corresponding mean efficiency rate is compared for all the algorithms. From the analysis it is found that the SSA outperformed all other algorithms with better efficiency rate, and after the SSA, the MOA has shown improved performance. So, these two algorithms have been chosen to perform feature selection. The efficiency rate is presented as follows:

$$\eta_{rate} = \frac{\sum_{i}^{N} \eta_i}{N} \times 100 \quad (2.26)$$

where η_{rate} is the better efficiency of a given algorithm over other algorithm, and η_i is the efficiency of ith experimental run. N is the number of times the given experimentation is made. The developed HMSA optimization algorithm is presented in Table 2.1, and the flowchart is presented in Figure 2.4.

TABLE 2.1
Algorithm of the Proposed HMSA

Input parameters
$X = (x_1, x_2, x_3, \ldots, x_n)$ – population vector
Y = minf(MSE) – fitness function
Output parameter
  Best fitness value
Set Parameters
  N – 500 // maximum number of iterations
  η_{rate} – 92 //Minimum Efficiency to be obtained for SSA
  P_{wh} (minimum) – 0.2
  P_{wh} (maximum) – 1.0
  Exploitation accuracy (ea) – 6.0
  Convergence criteria – 10^{-6}
  α – 0.2, β – 0.7, δ – 0.3 and λ – 0.4
  Pop_Max = 50
Invoke SSA
  Generate spiders around 'FF_Min' range domain attained from SSA approach
  Assign genders for spiders (65%–90% for female and remaining males)
WHILE ((Convergence criterion is not satisfied) DO
  Evaluate fitness of each spider in the population 'minf
  Define: Best_spider in the swarm, best female_spider, closest_spider
  IF (spider==female)
  IF (female_spider ==near to source)

$$X_m(t+1) = X_m(t) + \begin{pmatrix} \alpha.Vbr\_h_m.\left(F_c - X_m(t)\right) + \beta.Vbr\_b_m.\left(F_b - X_m(t)\right) \\ +\delta.\left(\lambda - \frac{1}{2}\right) \end{pmatrix}$$

ELSE IF (female_spider ==move away from source)

$$X_m(t+1) = X_m(t) - \begin{pmatrix} \alpha.Vbr\_h_m.\left(F_c - X_m(t)\right) + \beta.Vbr\_b_m.\left(F_b - X_m(t)\right) \\ +\delta.\left(\lambda - \frac{1}{2}\right) \end{pmatrix}$$

  END IF
ELSE IF (spider==male)
  IF (male_spider==non-dominant)

$$X_{m-ND}(t+1) = X_m(t) + \left(\alpha.Vbr\_f_m.\left(F_f - X_m(t)\right) + \delta.\left(\lambda - \frac{1}{2}\right)\right)$$

ELSE IF (male_spider==dominant)

$$X_{m-D}(t+1) = X_m(t) + \left(\alpha.\left(\frac{\sum_{n=1}^{N_m} \omega_n^k . W_{N_m} + n}{\sum_{n=1}^{N_m} W_{N_m}}\right) - X_m(t)\right)$$

TABLE 2.1 (Continued)
Algorithm of the Proposed HMSA

END IF
END IF
Create New spiders based on mating radius point of male and female radius
IF (new spiders possess better fitness than previous ones)
Replace worst spiders with new spiders
END IF
IF ($\eta_t < \eta_{rate}$) THEN
Discontinue SSA and invoke MOA with the Ybest of SSA
for all population do
evaluate minf() value
sort the particles from best finess to worst fitness value
Compute norm of solution set - Norm(Ui_i)
Update (x_{best_j})// best solution set
for all worst solution set 'i' do
compute P_{wh} and ρ
In_bhole = i
for all x_j do
rand1=random(0,1)
if ($rand1 < Norm(U_i)$) Then
selectIn_wthole by Roulette Wheel selection
S(In_bhole, j)=Sort(In_wthole, j)
end if
r3=rand(0,1)
if ($r3 < P_{wh}$) then
r2=rand(0,1)
r1=rand(0,1)
if (r2<0.5) then
Update the position of solution_universe with

$$x_i^j = x_{fit\_j} + \eta \times \left(\left(upr\_bd_j - low\_bd_j\right) \times r1 + low\_bd_j\right)$$

elseif ($r2 \geq 0.5$) then
Update the position of solution_universe with

$$x_i^j = x_{fit\_j} - \eta \times \left(\left(upr\_bd_j - low\_bd_j\right) \times r1 + low\_bd_j\right)$$

end if
end if
if ($r3 < P_{wh}$) then)

$$x_i^j = x_i^j$$

end for
end for
END WHILE
Return minimum fitness value and the corresponding coefficients.

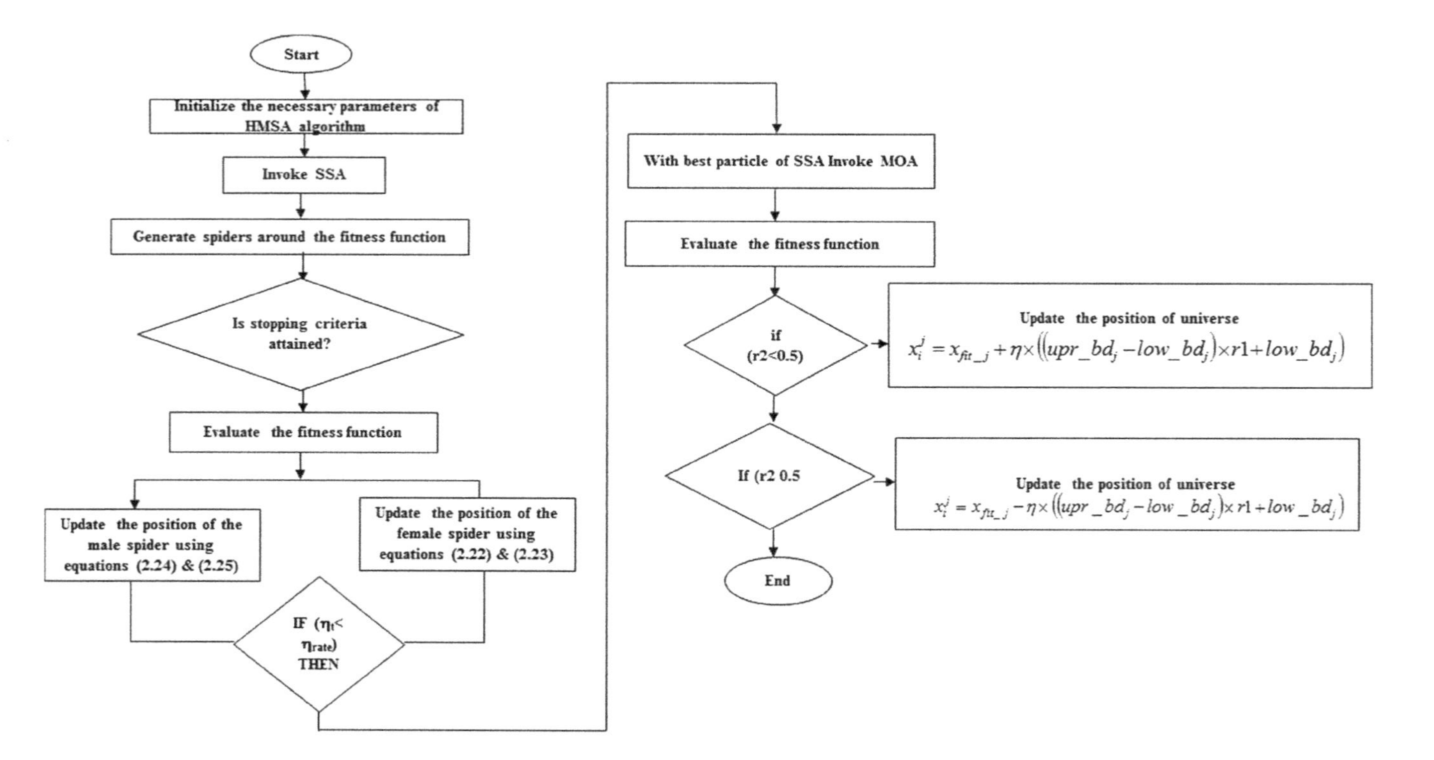

FIGURE 2.4 Flow chart of proposed HSSA.

2.5 EXPERIMENTAL ANALYSIS OF PROPOSED HYBRID FAULT DETECTION MODEL

The PV system is simulated in a MATLAB® environment for its healthy and faulty conditions. About 25,543 data sets were simulated under normal operating conditions, 2,234 data sets corresponding to short-circuit fault, 5,341 representing the open-circuit fault, 4,346 representing the shading condition, 2,093 corresponding to degradation fault, 1,184 representing the line-to-line fault, 4,094 corresponding to bridge fault, and 3,127 signifying the bypass diode fault—all are presented in Table 2.2.The entire data set is segregated into 70% for training purpose that comprises of 17,880 data sets representing the normal condition and 13,593 of abnormal condition and 30% for testing that consist of 7,663 of normal data and 5,826 of abnormal data. The classification performance of the proposed model is evaluated by means of the following performance metrics:

Accuracy: The quantity of accurate classification made by the algorithm.

$$Accuracy = \frac{TP + TN}{TP + FP + TN + FN} \tag{2.27}$$

Precision: The degree of the classifier to identify what percentage of the relevant classification is made by the proposed classifier algorithm.

$$Precision = \frac{TP}{TP + FP} \tag{2.28}$$

Recall/True Positive Rate (Sensitivity): The extent of true positive is identified as positive by the proposed algorithm.

$$Recall\ or\ TPR = \frac{TP}{TP + FN} \tag{2.29}$$

TABLE 2.2
The Training and the Testing Data Set

Condition	Training	Testing	Number of Data
Normal	17,880	7,663	25,543
Short circuit	1,570	664	2,234
Open circuit	3,739	1,602	5,341
Shading condition	3,042	1,304	4,346
Degradation fault	1,465	628	2,093
Line to line fault	829	355	1,184
Bridge fault	2,866	1,228	4,094
Bypass fault	2,189	938	3,127

True Negative Rate (Specificity): The extent of true negative is identified as negative by the proposed algorithm.

$$Specificity\ or\ TNR = \frac{TN}{FP + TN} \tag{2.30}$$

F1 Score: The Harmonic mean of precision and recall.

$$F1\,Score = \frac{2 * precision * recall}{(precision + recall)} \tag{2.31}$$

2.6 RESULT AND DISCUSSION

In this chapter, the mathematical modeling of the PV module is presented, and the data set corresponding to healthy and faulty condition is simulated and employed as input for the developed fault diagnosis strategy. The probability density model is developed in this article, to perform accurate fault classification. To enhance the performance of the proposed model, the feature selection is made by the proposed novel hybrid Social Spider Optimization algorithm on combining the best characteristics of MOA and SSA. Generally, the swarm intelligence algorithm suffers from certain issues like local stagnation problem, global minima occurrences, instability occurrences of network architecture, delayed convergence, premature convergence, and so on. So, in order to attain a better trade-off between the exploration and exploitation ability of the proposed hybrid swarm intelligence technique, the exploration ability of MOA strategy is combined with the exploitation ability of SSA. The efficiency rates of various optimization algorithms are computed and their performances compared with each other. Among all the algorithms considered to perform feature selection, the MOA and SSA reported better results than that of other algorithms, as reported in Table 2.3. So these two algorithms are combined as a hybrid optimization strategy

TABLE 2.3
η_{rate} of Proposed HMSA with Various Optimization Algorithms

Data Set	η of Algorithm Under Study in %											
	GA	PSO	BBO	FPA	GWO	BAT	FA	CS	TLBO	MFO	SSA	MOA
#1	80	89	79	81	85	84	86	84	85	84	89	90
#2	83	84	85	82	82	83	88	82	79	82	88	94
#3	88	86	83	83	87	85	84	81	82	86	90	92
#4	90	83	81	84	85	81	85	80	84	82	86	93
#5	87	90	80	86	86	86	82	84	81	84	87	92
#6	83	92	86	83	85	84	80	85	85	83	87	94
#7	82	80	78	81	83	86	81	85	83	86	86	89
#8	87	84	80	80	86	88	83	83	82	80	87	94
#9	89	89	82	79	84	83	80	82	81	83	91	90
#10	90	92	84	90	88	89	87	86	85	86	89	94
η_{rate}	**85.9**	**86.9**	**81.8**	**82.9**	**85.1**	**84.9**	**83.6**	**83.2**	**82.7**	**83.6**	**88**	**92.2**

TABLE 2.4
Performance Metric Results of the Proposed PNN Strategy

Model Under Study	Accuracy	Precision	Sensitivity	Specificity	F1 Score
PNN-SSA	88.325	94.366	86.383	91.529	90.198
PNN-MOA	89.097	94.615	87.296	91.998	90.808
Proposed PNN-HMSA	**99.459**	**99.189**	**99.869**	**98.938**	**99.523**

TABLE 2.5
Comparative Analysis with Existing Works of Literature

Model Under Study	Fault Identification Accuracy						
	Short Circuit	Open Circuit	Shading Condition	Degradation Fault	Line to Line Fault	Bridge Fault	Bypass Fault
PNN (Zhao et al. 2020)	92.56	93.09	92.42	92.09	93.03	92.96	93.27
RBFNN (Hussain et al. 2020)	94.98	94.54	95.48	95.98	94.87	94.46	95.93
EWMA (Garoudja et al. 2017)	89.98	87.93	90.98	88.67	89.65	89.49	90.00
Proposed PNN-HMSA	**99.23**	**99.43**	**99.19**	**98.99**	**99.34**	**99.04**	**99.28**

to perform effective attribute selection. Now, the attributes selected by the proposed hybrid optimization algorithm feed into the proposed probabilistic neural network model, and the classification is performed. The obtained results are presented in Table 2.4; here the response of three main techniques are reported. The performance metric values of the proposed PNN combined with MOA, SSA, and the hybrid MSSA are tabulated and their performances are investigated and presented below.

The performance of the proposed hybrid MSSA achieved 10.32% improved accuracy vs. the PNN model feed with features selected by MOA algorithm. The precision was 4.57% better than the PNN-MOA strategy, the sensitivity attained was 12.57% better than the PNN-MOA algorithm, the specificity was 6.94% better than the PNN-MOA, and the F1 score was 8.72% better than the PNN-MOA algorithm. On comparing with the performance of the PNN-SSA algorithm, the MSSA achieved 11.13% better accuracy, 4.823% improved precision, 13.486% better sensitivity, 6.9% improved specificity, and the F1 Score is improvised to the value of 8.72%.

The performance of the proposed fault diagnosis strategy compared with existing works of literature such as PNN (Zhao et al. 2020), RBFNN (Hussain et al. 2020), and EWMA (Garoudja et al. 2017) is presented in Table 2.5. Various fault scenarios such as short-circuit fault, open-circuit fault, shading fault, degradation fault, L-L fault, bridge fault, and bypass fault are identified and the corresponding classification accuracy is reported in the table. The short-circuit fault identification accuracy is 92.56%

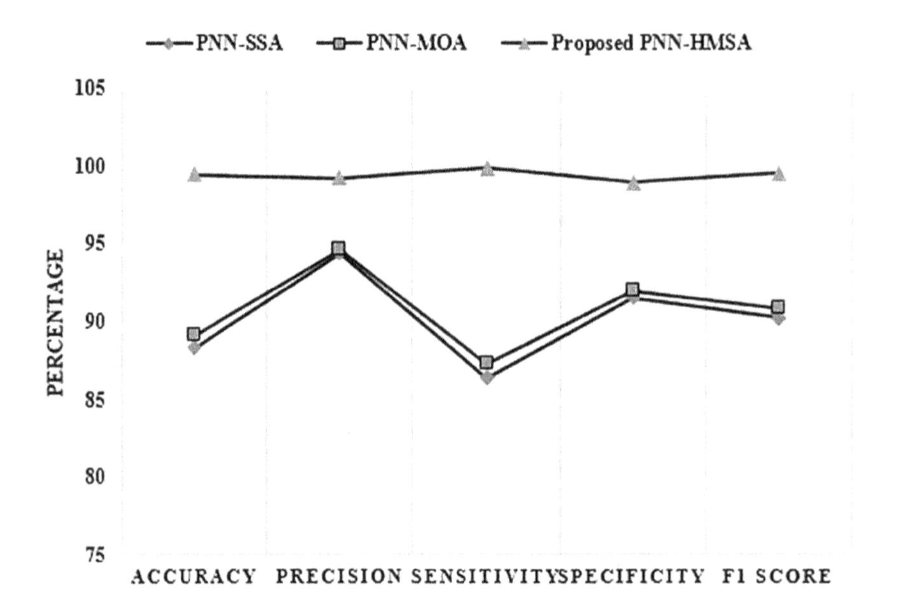

FIGURE 2.5 Classification performance of the proposed model.

for PNN, 94.98% for RBFNN, 89.98% for EWMA, and the proposed PNN-HMSA reported the short-circuit classification accuracy of 99.23%. The classification accuracy of open-circuit fault is 93.09% for PNN, 94.54 for RBFNN, 87.93 for EWMA, and 99.43% for proposed PNN-HMSA. For the shading fault scenario, the classification accuracy is reported as 92.42% for PNN, 95.48% for RBFNN, 90.98% for EWMA, and 99.19 % for the proposed hybrid PNN-HMSA. During degradation fault classification, the PNN performed with 92.09% of accuracy, 95.98% for RBFNN, 88.67 for EWMA, and the proposed PNN-HMSA outperformed with 98.99%. The line-to-line fault detection accuracy of 93.03% was reported for PNN, 94.87 attained for RBFNN, 89.65 %for EWMA, and 99.34% reported for PNN-HMSA. The performance of bridge fault detection is presented as 92.96% of classification accuracy for PNN, 94.46% of RBFNN, 89.49% for EWMA, and 99.28% of accuracy achieved for the proposed model. The classification accuracy for bypass fault is reported for PNN to be 93.27%, 95.93% for RBFNN, 90.00% for EWMA, and 99.28% for PNN-HMSA. On comparing the performance of other existing strategies, the proposed PNN-HMSA model outperformed with better classification metric results for all the fault scenarios considered in this study (Figure 2.1, Figure 2.4, Figure 2.5).

2.7 CONCLUSION

In this chapter a novel fault diagnosis strategy has been developed to identify the type of fault in a photovoltaic system. The PV faults such as open-circuit fault, shading condition, degradation fault, line-to-line fault, bridge fault, and bypass diode fault were diagnosed by a PNN classifier algorithm. To improve the performance of the proposed strategy, a hybrid feature selection algorithm has been developed based on combining the best features of Multiverse Optimization Algorithm (MOA) and Social Spider Optimization Algorithm (SSA). The proposed hybrid strategy has significantly improved the performance of the model by reducing the data dimension and data redundancy. The performance of the proposed strategies was compared with each other and compared with existing works of literature. Based on the experimental analysis it was demonstrated that the performance of proposed PNN with HMSA algorithm outperformed all other techniques with better classification metric results.

REFERENCES

Bendary, Ahmed F., Almoataz Y. Abdelaziz, Mohamed M. Ismail, Karar Mahmoud, Matti Lehtonen, and Mohamed M.F. Darwish. "Proposed ANFIS based approach for fault tracking, detection, clearing and rearrangement for photovoltaic system." *Sensors* 21, no. 7 (2021): 2269.

Braun, Henry, Santoshi T. Buddha, Venkatachalam Krishnan, Andreas Spanias, Cihan Tepedelenlioglu, Ted Yeider, and Toru Takehara. "Signal processing for fault detection in photovoltaic arrays." In *2012 IEEE International Conference on Acoustics, Speech and Signal Processing (ICASSP)*, pp.1681–1684. IEEE, 2012.

Cai, Yuqiao, Peijie Lin, Yaohai Lin, Qiao Zheng, Shuying Cheng, Zhicong Chen, and Lijun Wu. "Online photovoltaic fault detection method based on data stream clustering." In *IOP Conference Series: Earth and Environmental Science*, 431, no. 1, p. 012060. IOP Publishing, 2020.

Clark, Jhon W. "Probabilistic neural networks." *Evolution, Learning and Cognition* 3, no. 1 (1988): 129–180.

Cuevas, Erik, Miguel Cienfuegos, Daniel Zaldívar, and Marco Pérez-Cisneros. "A swarm optimization algorithm inspired in the behavior of the social-spider." *Expert Systems with Applications* 40, no. 16 (2013): 6374–6384.

Dhimish, Mahmoud, and Violeta Holmes. "Fault detection algorithm for grid-connected photovoltaic plants." *Solar Energy* 137 (2016): 236–245.

Dhimish, Mahmoud, Violeta Holmes, Bruce Mehrdadi, and Mark Dales. "Multi-layer photovoltaic fault detection algorithm." *High Voltage* 2, no. 4 (2017a): 244–252.

Dhimish, Mahmoud, Violeta Holmes, and Mark Dales. "Parallel fault detection algorithm for grid-connected photovoltaic plants." *Renewable Energy* 113 (2017b): 94–111.

Eskandari, Aref, Jafar Milimonfared, and Mohammadreza Aghaei. "Line-line fault detection and classification for photovoltaic systems using ensemble learning model based on IV characteristics." *Solar Energy* 211 (2020): 354–365.

Garoudja, Elyes, Fouzi Harrou, Ying Sun, Kamel Kara, Aissa Chouder, and Santiago Silvestre. "Statistical fault detection in photovoltaic systems." *Solar Energy* 150 (2017): 485–499.

Hajji, Mansour, Mohamed-Faouzi Harkat, Abdelmalek Kouadri, Kamaleldin Abodayeh, Majdi Mansouri, Hazem Nounou, and Mohamed Nounou. "Multivariate feature extraction based supervised machine learning for fault detection and diagnosis in photovoltaic systems." *European Journal of Control* 59 (2020): 313–321.

Hussain, Muhammed, Mahmoud Dhimish, Sofya Titarenko, and Peter Mather. "Artificial neural network based photovoltaic fault detection algorithm integrating two bi-directional input parameters." *Renewable Energy* 155 (2020): 1272–1292.

Kavi, Moses, Yateendra Mishra, and Mahinda Vilathgamuwa. "DC Arc fault detection for grid-connected large-scale photovoltaic systems." *IEEE Journal of Photovoltaics* 10, no. 5 (2020): 1489–1502.

Khelil, Chérifa Kara Mostefa, Badia Amrouche, AbouSoufiane Benyoucef, Kamel Kara, and Aissa Chouder. "New Intelligent Fault Diagnosis (IFD) approach for grid-connected photovoltaic systems." *Energy* 211 (2020): 118591.

Kim, Il-Song. "On-line fault detection algorithm of a photovoltaic system using wavelet transform." *Solar Energy* 126 (2016): 137–145.

Lazzaretti, André Eugênio, Clayton Hilgemberg da Costa, Marcelo Paludetto Rodrigues, Guilherme Dan Yamada, Gilberto Lexinoski, Guilherme Luiz Moritz, Elder Oroski et al. "A monitoring system for online fault detection and classification in photovoltaic plants." *Sensors* 20, no. 17 (2020): 4688.

Li, B., Claude Delpha, D. Diallo, and A. Migan-Dubois. "Application of Artificial Neural Networks to photovoltaic fault detection and diagnosis: A review." *Renewable and Sustainable Energy Reviews* 138 (2020): 110512.

Madeti, Siva Ramakrishna, and S. N. Singh. "Modeling of PV system based on experimental data for fault detection using kNN method." *Solar Energy* 173 (2018): 139–151.

Mellit, Adel, and Soteris Kalogirou. "Artificial intelligence and internet of things to improve efficacy of diagnosis and remote sensing of solar photovoltaic systems: Challenges, recommendations and future directions." *Renewable and Sustainable Energy Reviews* 143 (2021): 110889.

Qureshi, Faheem A., Zahoor Uddin, M. Bilal Satti, and Muhammad Ali. "ICA-based solar photovoltaic fault diagnosis." *International Transactions on Electrical Energy Systems* 30, no. 8 (2020): e12456.

Rao, Sunil, Andreas Spanias, and Cihan Tepedelenlioglu. "Solar array fault detection using neural networks." In *2019 IEEE International Conference on Industrial Cyber Physical Systems (ICPS)*, pp. 196–200. IEEE, 2019.

Triki-Lahiani, Asma, Afef Bennani-Ben Abdelghani, and Ilhem Slama-Belkhodja. "Fault detection and monitoring systems for photovoltaic installations: A review." *Renewable and Sustainable Energy Reviews* 82 (2018): 2680–2692.

Zhang, Jingwei, Yongjie Liu, Yuanliang Li, Kun Ding, Li Feng, Xihui Chen, Xiang Chen, and Jiabing Wu."A reinforcement learning based approach for on-line adaptive parameter extraction of photovoltaic array models." *Energy Conversion and Management* 214 (2020): 112875.

Zhao, Jian, Qian Sun, Ning Zhou, Hao Liu, and Haizheng Wang. "A photovoltaic array fault diagnosis method considering the photovoltaic output deviation characteristics." *International Journal of Photoenergy* 2020 (2020).

Zhu, Honglu, Sayed Ahmed Zaki Ahmed, Mohammed Ahmed Alfakih, Mohamed Abdelkarim Abdelbaky, Ahmed Rabee Sayed, and Mubaarak Abdulrahman Abdu Saif. "Photovoltaic failure diagnosis using sequential probabilistic neural network model." *IEEE Access* 8 (2020): 220507–220522.

3 A Remote Diagnosis Using Variable Fractional Order with Reinforcement Controller for Solar-MPPT Intelligent System

Johny Renoald Albert and Thenmalar Kaliannan
Vivekanandha College of Engineering for Women, Namakkal, India

Gopinath Singaram
Annasaheb Dange College of Engineering and Technology, Sangli, India

Fantin Irudaya Raj Edward Sehar
Dr. Sivanthi Aditanar College of Engineering, Tiruchendur, India

Madhumathi Periasamy
Anna University, Chennai, India

Selvakumar Kuppusamy
SRM Institute of Science and Technology, Chennai, India

CONTENTS

DOI: 10.1201/9781003202288-3

LEARNING OUTCOME

i. To assess how Reinforces Learning Approach (RLA) of fractional order and Deep Q-Learning (DQL) pattern can be implanted to operate effectively in a different scenario.
ii. To study the real data collected from the Climate Change Information System (CCIS) in Tamilnadu.
iii. To discover how the Fractional Order with Reinforcement Controller (FORC) helps reduce the tracking time, oscillation around the maximum, and minimized harmonic distortions.

3.1 INTRODUCTION

In various climatic conditions, the development of a Maximum Power Point (MPP) controller is an integral part of the photovoltaic system to make certain continuous energy delivery in dynamic load conditions. Regarding the changes in environmental conditions and their parametric variation, the most difficult factor here to design a model that can monitor the maximum power delivery. The model conceived in this work addresses both challenges because it Reinforces Learning Approach (RLA) of fractional order. The Deep Q-Learning (DQL) pattern can be trained into the model, which can then be implanted to operate effectively in a different scenario. The combination of Fractional Order with Reinforcement Controller (FORC) helps reduce the tracking time, oscillation around the maximum, and total harmonic distortions. In this context, the model is well tested and has achieved the desired results under standard test conditions. The design was checked on the real data collected from the Climate Change Information System (CCIS) in Tamilnadu. The MATLAB® results of Fractional DQN-MPP tracking approach are to give better outcomes with a comparison of intelligent controller DQN MPP, Fuzzy MPP, and P&O MPP even in lower photovoltaic (PV) irradiance conditions.

Shunmugham Vanaja et al. (2021) have compared solar energy direct power generation with other sustainable resources. Johny Renoald Albert (2020), Abhishek Gautam et al. (2020), Priyanka Roy et al. (2020), Jeffrey M et al. (2021), and Goswami Y et al. (2021) have described sun irradiation captured through a variety of technologies, which are still being developed, as well as solar heating, solar thermal energy, solar architecture, molten salt power plants, and artificial photosynthesis. Sandro Nizetic et al. (2019) have discussed the reduction of global warming and economic efficient society through the use of renewable energy sources and calculate the energy efficiency in three major areas: green buildings power consumption, solar energy utilization efficiency waste elimination, and smart cities with the Internet of Things (IoT). Manoja Kumar Behera and Lalit Chandra Saikia (2020) have proposed a recent algorithm combination of extreme learning machine variable uses the Steepest Gradient Ascent (SGA) for maximum power extracting in a solar PV system. SGA is combined with Salp Swarm Algorithm (SSA), and they use Proportional Integral (PI)-Fractional Order Integral (FOI) cascade controller to calculate better tracking of solar irradiance and tracking effectiveness in real-time climate conditions to achieve greater accuracy.

Albert and Stonier (2020) and Benlahbib et al. (2018) have developed a solar-MPP system with three variables to calculate data in real time according to weather conditions in less setting time and maximum oscillation point, called Fuzzy Logic Controller (FLC). They then reviewed optimization of power system problems on smart grid concepts. The review concluded that integration of distributed sources, demand feedback, and electric vehicle on the electricity board is poor. It can be resolved by some advanced computational tools like computational intelligence and new optimization techniques for easy analysis of a problem and better quality of power. Minai and Malik (2021) analyzed the survey on MPP technique regarding the issues of solar PV system around 30–40% of solar irradiance used for electricity production. In this study, harmonics in a PV system were reduced by using an MPP algorithm and power electronic converters. Pakkiraiah and Sukumar (2016) applied a test to a PV system for resolution of complex problems in operating point depending on climate condition. They developed a PV emulator based on a logarithmic of ideal single diode model (ISDM) to achieve low-cost, simple, and effective device to account for climate-changing conditions. The solar panel is calculated to emulate with programmable power supply designed with an ability to recharge 12V, 7 Ah batteries.

Merenda, Massimo et al. (2019) and Saadullah Khan et al. (2018) reviewed the electric vehicles (EVs) in term of different aspects such as sources in fossil fuel–based generation system. They have comprehensively analyzed solar PV EV charging systems and determined their time and cost effectiveness. Mazzeo et al. (2020) proposed a new energy-conserving environmental multi-criteria decision-making in different optimization aspects in a cascade renewable system. Based on the new dimension, fewer indicators were used to compare the PV vs. wind, PV vs. battery, and wind vs. battery power generation. The results showed improvements in cost optimization, emission reduction, specific load values, and increase in battery capacity. Rezkallah et al. (2015) discussed the real-time Hardware in the Loop (HIL) implementation of sliding mode control in a PV system with the dump load under different weather conditions. The Battery Storage System (BESS) was protected by a Currently Controlled Voltage Source Converter (CC-VSC). Stability was analyzed, along with load perturbation over variation of the climatic condition. The energy integration of buildings and transportations is strong under all weather environments, according to Yuekuan Zhou et al. (2019). Capacity expansion in the power fluctuation of the electric grid is retrieved from the sources of low efficiency of heat. For monitoring and reducing expense and pollution, various controls are used.

Nkambule et al. (2021) analyzed PV panels in terms tracking of maximum power. They used three different MPP techniques for analyzing solar irradiation under different weather conditions. They concluded that the most accurate analysis is achieved from the review of various parameters such as convergence speed, accuracy, efficiency, system reliability, and cost of the system. In addition to that, there are six different MPP techniques that are implemented on a bio-inspired algorithm that gives better results with PV systems irradiances. Ying-Yi Hong et al. (2018) focused on the robust design of MPP using the Taguchi method for a stand-alone PV system. Orthogonal experiments were taken for reducing the full factorial experimental design under the Taguchi method. This was based on maximum power point tracking for isolation, tilt angle, and local resistance. The OP5600 real-time digital simulator uses the irradiance and temperature for better accuracy of power.

Bubalo et al. (2021) concluded that the renewable sources like PV system have advantages of low-cost maintenance compared to non-renewable sources when the Z-Source inverter is used on the FLC Algorithm, achieving better results for a solar array on chroma 62 150H – 600s/1000s. The Perturb & Observe (P&O) algorithm and FLCA are implemented for better outcomes of solar irradiance. Thenmalar Kaliannan et al. (2021) and Dhivya et al. (2017) suggested a resolution to the problems of electricity under particular weather conditions. They used the recent MPP, known as Transfer Reinforcement learning (TRL) with new optimization techniques, to compare with the incremental conductance (INC) model. They analyzed the outcomes from step-up-step, step-change in solar with temperature variation, and radiation. The reinforcement learning implement in MPP simplifies the issue because it is a free solution that eliminates the impact of test parameters on model architecture.

If the model has been developed, it can be used in a variety of environments with incredible precision without the need to remodel or retune the parameters. In the RLA algorithm, Q-learning is a model-free solution. Here, the agent is led to a new course of action based on previous experience. This algorithm calculates the state-action value function for a goal policy that selects the highest-valued action. This condition is under control until the search area is reduced in size. When the search space has high dimensionality, it is essential to store millions of records in a table in the program memory. This flaw is overcome by using the principle of Deep Q-Learning, in which a neural network chooses the behavior. Noninteger-based fractional order control methods are usually favored over integer-based control methods due to their accuracy and more discrete space. It also provides a more modular architecture, which has improved performance robustness. Furthermore, in traditional methods of tracking MPP, a fixed amplitude perturbation is introduced, but oscillations are created due to a discrepancy between change in amplitude and tracking speed phase scale, resulting in reduced performance. With the support of the fractional solution, this flaw is effectively addressed. The rate of evaporation is greater in the summer than in the winter, as shown in Figure 3.1, and the state has opaque rain-bearing clouds in December, November, and October. In certain cases, it has been found that the conventional design struggles to detect the MPP in real-time changing environmental conditions.

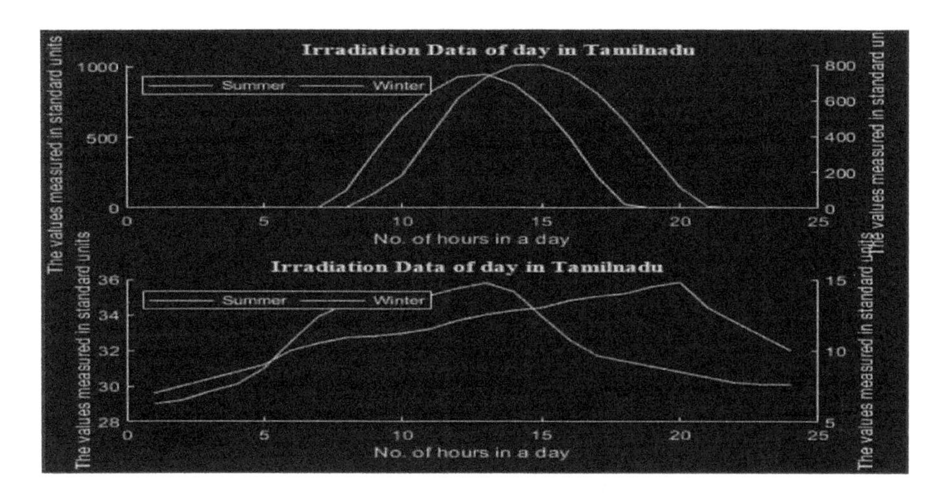

FIGURE 3.1 Climate change in TN state, summer and winter.

Design and development of test using real-world evidence are presented. This information comes from a CCIS tool. Tamilnadu (TN) will be the focus of the investigation. The rationale for collecting this information is that our state's capital is one of the most heavily populated cities, and it experiences severe power shortages during the year. Motivated by a research gap and the points made earlier in the chapter, we developed a novel fractional-order MPP control algorithm with DQN in reinforcement learning for a PV array under variable load for partial shading conditions. The monitoring of MPP is controlled by the RLA agent, which is fed with seven separate observations for the learning phase, including PV capacity, voltage, PV generated power diversion with PV's desired capability, integral of power diversion, coupling voltage, and divergence per unit time with reference coupling DC.

The agent's tracking speed is then indirectly helped by the performance of the fractional order with error and derivative error words, making the process more effective. It is also checked on real data for some specific places in TN. The following sections are organized as follows: Section 3.2: Solar performance in TN. The developed PV framework is introduced in Section 3.3. Intelligent controller implementations are then discussed in Section 3.4. Section 3.5 outlines the possibilities of the planned process in the proposed implementation of Fractional DQN with RLA, as well as closing remarks, which include the proposed algorithm with RLA formulation for MPP Control and Deep RLA tracking problem.

3.2 SOLAR PERFORMANCE IN TAMILNADU

India accounts for 2.4% of the world's gross land area. The maximum length of the mainland from south to north is around 3,214 km. The country's latitudinal and longitudinal extents are almost identical in degrees, i.e., around 30°. Rajasthan has been believed long ago to be India's sunniest state, but the Indian Space Research Organization (ISRO) mapping of solar energy hot spots revealed that Gujarat receives the highest levels of solar radiation, making it the country's best place to construct solar power plants. TN is India's eleventh-largest state, with a total area of 130,058 km^2 (50,216 sq mi). It borders Kerala in the west, Karnataka in the northwest, and Andhra Pradesh in the north. Chennai gets the most solar energy in December from TN, and was the hot zone in 2020. Temperature extremes would almost definitely increase as global mean temperatures rise (IPCC 2014). TN Climate Change Information System (CCIS) offers science-based climate change information that increases user knowledge and understanding of climate change at the state level. It provides authoritatively observed and projected temporal and spatial information throughout the district.

TN has reported that by 2023, it aims to have 9,000 MW of solar power installed. Two current installations have a 235 MW capacity and two others have a 600 MW capacity. The complex created 1,194 MU in 2019–2020. The CCIS system analyses solar efficiency, with the most significant climatic components including wind, pressure, temperature, humidity, clouds, and precipitation. TN has a tropical climate change measured by an Application Programming Interface (API) tool kit that produces real-time, historical, and forecast estimates of the available resources around the globe. It also provides actual and forecast solar irradiance and power data, globally using satellites and surface measurements. Two variables affect the climatic conditions in TN: the apparent path of the irradiation sun and the monsoonal rain-bearing winds. The sun's vertical rays strike the state twice a year. The state's relative humidity is greater in the winter than it is in the summer. In May, the average humidity is around 68%, while it is 82% in January. The data for the year 2020 was taken and the output PV power and the other standard values were measured for a day. The simulation (MATLAB 2020a) results for summer and winter testing under different environmental conditions are shown in Figure 3.2a and Figure 3.2b, respectively. Table 3.1 shows the different solar irradiation data in Tamilnadu, and Table 3.2 shows the region-wise seasonal average temperature.

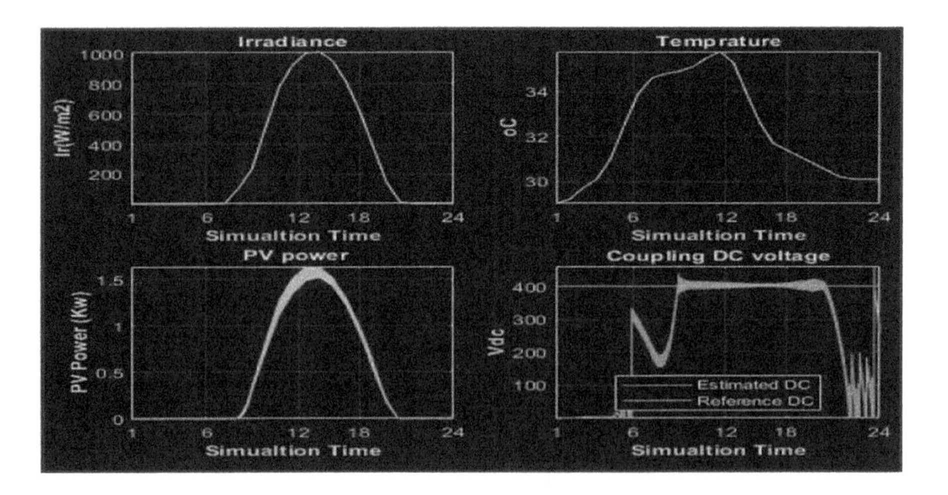

FIGURE 3.2 (a) PV power and DC coupling voltage for testing under summer data.

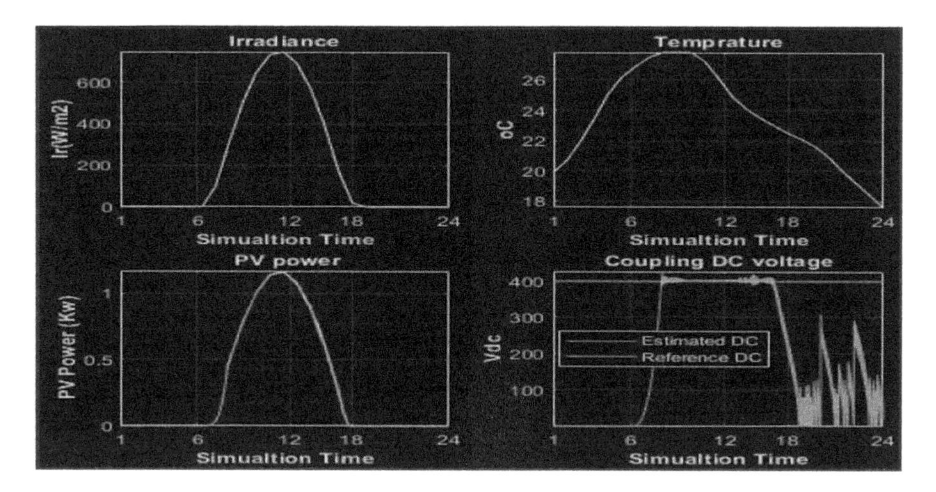

FIGURE 3.2 *(Continued)* (b) PV power and DC coupling voltage for testing under winder data.

TABLE 3.1
Solar Irradiation Data in Tamilnadu

District	Latitude	Longitude	Meteronorm Data-GHI	Nasa Data GHI	Variation	Variation %
Nilgiris	11.6167	76.5	5.45	4.97	0.48	8.81
Chennai	13.15	79.9167	5.5	5.11	0.39	7.09
Madurai	9.8	78.1	5.59	5.1	0.49	8.77
Tirchy	10.8	78.6833	5.55	5.2	0.35	6.31
Salem	11.65	78.1667	5.58	5.19	0.39	6.99
Vellore	12.9167	79.1333	5.58	5.13	0.45	8.06
Coimbatore	11.2667	76.9833	5.52	4.97	0.55	9.96

TABLE 3.2
Area-Wise Seasonal Average Temperatures

Geographical Locations	Weather Footage Stations	Temperature in Celsius		
		Summer	Winter	Rainy
Plateau	Covai	32	26	21
Interior Plains	Vellore	42	21	22
Hilly regions	Kodaikanal	11	6	14
Inland regions	Tirchy	42	20	23
Coastal regions	Chennai	40	22	25

TABLE 3.3
Parameter for the Solar PV Array SPR-306E-WHT-D

Parameters	Units
No. of PV array series connected module	9
No. of PV array parallel connected module	250
Voc PV array	65.2V
Isc PV array	5.95A
Maximum power standard operating conditions	305W
Irradiation range I_r	680–1500
Voltage at the maximum power point V_{PV}	55.7V
Current at the maximum power point I_{PV}	5.59A
Ref. DC coupling voltage	400V

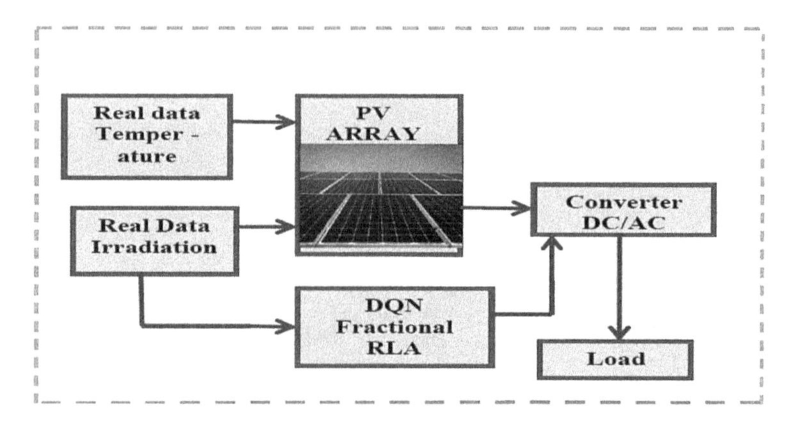

FIGURE 3.3 The DQN with the fractional design of MPP control.

3.3 PROPOSED PV ARRAY PARAMETERS

The proposed fractional MPP scheme is tested when powered from the SPR 306E-Watt sun module. The module is tested under the standard conditions and for the real data set. The parameters employed in a PV array simulation are presented in Table 3.3. The performance of the PV cell taken for DQN with the fractional design of MPP control model is showcased in Figure 3.3. This figure shows the intermediate fractional block configuration.

3.4 INTELLIGENT CONTROLLER IMPLEMENTATION

Fractional order for MPP-PV array's functionality is usually performed under changing atmospheric conditions affected by parameters such as radiation and heat. They have an overall effect on the PV range. Therefore, for modeling the PV module, it is essential to consider them. There are methods for estimating parameters, one of which is MPP. V is the equivalent tension and current of every PV curve's highest point. When considering the PV power (P_{PV}) derivative concerning PV voltage (V_{PV}), it can approximate for a certain environmental situation the parameters for a difference in MPP.

$$[\![dP]\!]pv/[\![dV]\!]pv\Big|(V_m,I_m) = I_m + [\![dI]\!]\,pv/[\![dV]\!]\,pv\Big|(V_m,I_m)\times V_m \tag{3.1}$$

PV systems are identical with processes such as diffusion, solar radiation, temperature, and electric driving. The fractional-order solution should illustrate this.

3.4.1 Fractional-Order Differentiator

The architecture of the FO is usually based on fractional derivative concepts. All sorts of variants, such as lighting and electrical flux, can be expressed by FO in the range of $0 < \alpha \sim 1$, derivative formulation.

3.4.2 Fractional-Order Control MPP Algorithm

The operation point is moved to the peak in the P-V curve to extract from the source the optimum power output. A fractional α (alpha) factor is used to achieve the peak quicker, enabling the planned model to have a vector domain for expansion and constriction.

If the input variable is extended, the built model has a broad monitoring stage to easily hit MPP. Equation (3.2) in the presence of α as updated.

$$(d\wedge\alpha P)/[\![dV]\!]\wedge\alpha = [\![\lim]\!](\Delta V\to 0)[\![1/[\![\Delta V]\!]\wedge\alpha]\!]\times \Sigma(t=0)\wedge\infty \left[\!\!\left[\begin{matrix}(-1\wedge t\,r(\alpha+1))/(r(t+1)r(\alpha-t+1))\\ P(V-t\Delta V)\end{matrix}\right]\!\!\right] \tag{3.2}$$

The power output of the PV array is $\alpha > 0$, $P\,(V - t\,\Delta V)$ at the moment of t time. The power and voltage shifts are approximated by the calculation in fractional order. Equation (3.3) is approximated

$$(d\wedge\alpha P)/[\![dV]\!]\wedge\alpha = [\![\lim]\!]\top(\Delta V\to 0)\,[\![(P-\alpha P_0)/[\![(V-V_0)]\!]\wedge\alpha]\!] \tag{3.3}$$

The P and P_0 are the power measured at instant t and t–1, respectively, and same for the voltage (V) and output voltage (V_0). The incremental power as approximated using the fractional-order is $d^\alpha P \cong P - \alpha P_0$, and the incremental change in the voltage is $dV^\alpha \cong (V - V_0)^\alpha$. The change in voltage and power in unit time is considered as error $E_r(t)$. The change in power is $\delta P(t)$. These values are calculated in Equations (3.4) and Equation (3.5), respectively.

$$E_r\,t = (P-P_0)/(V-V_0) \tag{3.4}$$

$$\delta P(t) = P - P_0 \tag{3.5}$$

This affects the input domain of the incentive function developed for the reinforcement learning agent and the α-fractional component falls below $0 < \alpha$ to 1. As $\alpha = 1$ is entered by the other inputs, the input is effectively changed by the input pattern of the incentive. The smaller value of α increases $E_r(t)$, which ultimately results in less time to reach MPP by increasing phase duration, and prevents oscillations at the optimum stage. Rapid MPP monitoring and stability of MPP performance are often designed to minimize energy losses and increase device reliability.

3.5 PROPOSED IMPLEMENTATION OF FRACTIONAL DQN WITH RLA

In this proposed method, we suggest a fractional order in the RLA to address the issue of MPP in PV arrays. Without parametrical knowledge of the complex parameters of the model, RLA effectively resolves this problem. The algorithm attempts to represent the machine comparison based on the DQN model.

3.5.1 RLA Formulation for MPP Control

Strengthening learning includes four major components. These are state-space X, compensation r, likelihood of transformation p, and the space of operation U. Every operation in the case of the MPP check is a manipulation of any vector V in a certain amount of time. Each interaction with the agent and the environment results in the learning process, and the action u (t) as a result of this mechanism develops from the state of the agent to the next state action. The agent then achieves reinforcement known as a reward that measures the quality $x_t + 1$ or the phase selected by the agent. This is why this award is a "pin" toward the objective or desired solution that is feasible. The RLA approach is intended to find a satisfying optimum strategy.

$$J* = (\pi\text{^ max})J_\pi = (\pi\text{^ max})E_\pi\{r_t / x_t = x\} \tag{3.6}$$

Here J_π corresponds to the expected reward against the policy π. Consider a forgiven policy π, the expected cumulative reward $V^\pi(x)$ or the value function for a certain time interval is the function of x^π and is defined as $x\text{^}\pi = \{x_t\}(t = 1)\text{^}(t = n)$. These are the state values $k\text{^}\pi = \{k_t\}$ $(t = 1)\text{^}(t = n)$ which are the sequences of the agent taken based on the actions.

i. **State Space**
   The different states in every MPP regulation are created using awareness of MPP's movement on the PV curve under various environmental conditions. Present, control, and voltage play a role in the RLA technique. The state-space includes V_{PV}, I_{PV}, P_{PV}, ΔP_{PV}, $\int\Delta P_{PV}$, $\alpha_{e(t)}$, and the coupling point ΔV_{DC}. Here $\int\Delta V_{DC}$ controls the selection of duty cycle in the interval of [0, 1]. ΔP_{PV} defines the divergence of *PV* power from the desired to the generation capacity and ΔV_{DC} is the differencebetween the reference coupling voltage and measured V_{DC}.

ii. **Action Space**
The action space used to solve the MPP problem is usually discrete. This ensures a high degree of accuracy while still acting as an effective learning strategy that makes the process computationally effective. There is a given service cycle in the operation of the RL-MPP agent. The duty cycle Dint is picked in the range Dint = (0,1) by a sequence of acts. As a consequence, a matrix of 100 potential acts is generated.

iii. **Reward**
Four values—VDC, PPV, e(t), and exb—have been used to create the incentive room. Except for the exb, which denotes the excess bound conditions, each input is subjected to the threshold. The generation of negative reward is caused by violations in the input values, while the permissible value of input fetches a positive reward value. Figure 3.4 depicts the internal design of obtaining e(t).

3.5.2 Deep RLA Tracking Problem

To achieve the required accuracy and robustness, designing an RLA agent is a particularly necessary and crucial step. The most difficult problem here is dealing with the continuous space. It is possible to strike the correct balance between excessive discretion and inadequate discretion. This restricts RLA's functionality to a limited range, limiting its ability to perform in complex environments. As a result, a technique known as Deep Q-learning emerges, which has the advantage of being off-policy and model-free. DQN's layered form is shown in Figure 3.5. It is made up of different layers, such as a linked layer and Relu layers, that are stacked on top of each other. The other necessary parameters taken for designing the network are tabulated in Table 3.4 below. The algorithm discusses the proposed Deep RLA control algorithm in this section. The RLA agent is based on the DQN architecture. As a result,

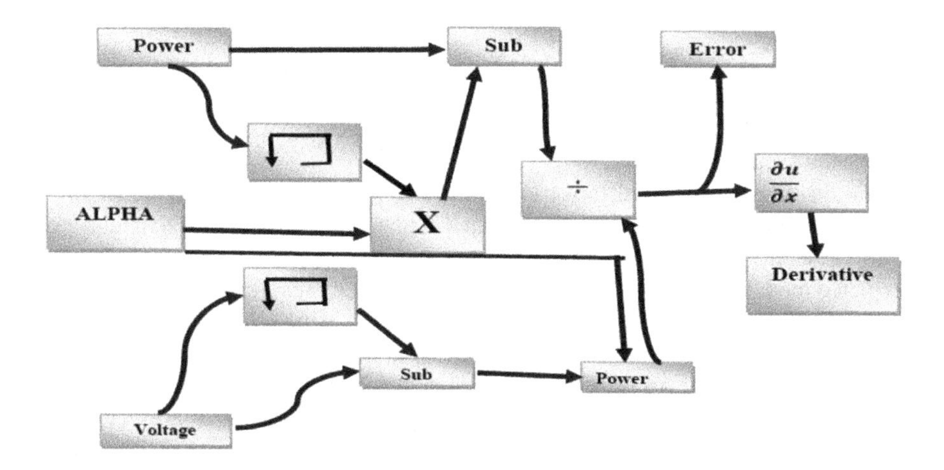

FIGURE 3.4 Schematic design of the fractional block.

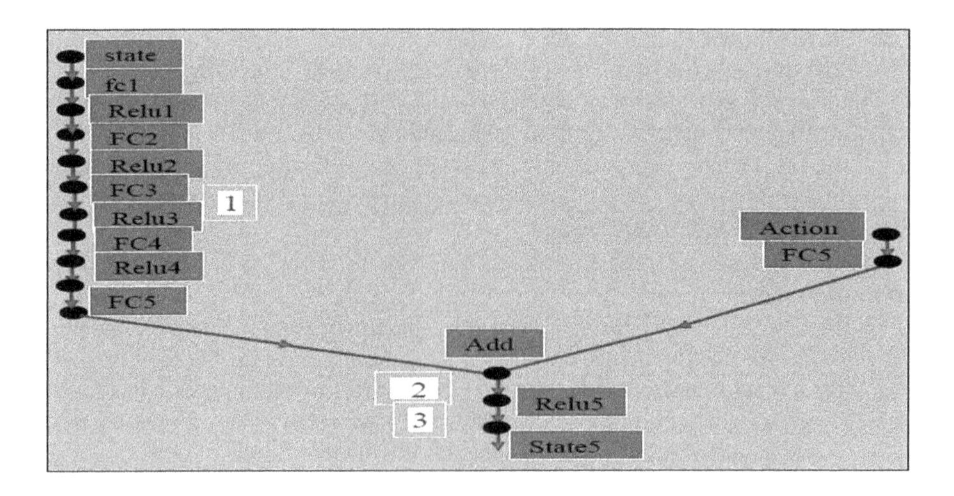

FIGURE 3.5 Internal structure of DQN employed for fractional MPP approach.

TABLE 3.4
The Parameter Setting for the Fractional DQN Network

Parameters	Value
Various Inputs	7
Relu Layers	5
Fully connected Layers	7
Number of Hidden neurons	100
Designed Discount Factor (γ)	0.99
Targeted Learning rate (α)	0.01
Maximum number of Episodes	300
Maximum step size	10

the agent is made up of neural networks that maximize policy and network performance. The algorithm begins by setting up the Q parameters for producing electricity by connecting PV arrays. The outcomes of this entire approach are described in the following section.

In this case, a fractional factor is used to transfer the error and derivative error terms as designed in the previous section. Then, as described in the technique section, these are placed in the network's state, action, and reward room. The above-depicted model DQN has been loaded. Following that, Q-learning is used to train the network. The network starts to fill up. Line 1 (Relu.3) is the origin of the main loop. The loop is programmed to run for many (M) episodes. The duty cycle's initial value is given, followed by all of the initial Dint states, observations, and the course of behavior. In each episode, line 17 leads to the inner loop x0 ut, which is executed for several time stages. In this case, the agent selects a (T) action from a pool of available actions in the environment (line number 18). As a consequence, a new state is created, and a reward is given based on the agent's behavior xt + 1, r. Line 2 (Relu5).

Finally, the three items are put into a buffer Line 3 (state5). The next step is to start training the agent if there are enough state transformations stored. The buffer is used to extract a minimal number (N) of random transformations. Q-learning is used to upgrade the network with these values. The cumulative number of training states for target networks has been completed and revised. The algorithm comes to a close on line 19, with all networks being conditioned and returning to the buffer.

3.6 ALGORITHM: FRACTIONAL ORDER DQN MPP

The performance of the PV cell taken for Simulink modeling is showcased in Figure 3.6. This curve is plotted for the standard temperature of 25°C under different irradiance conditions.

1. Connect solar PV array SPR-305-WHT-D (330 sun power)
2. Set the short-circuited current I_{sc} = 5.95A and open-circuited voltage V_{oc}= 65.2V
3. Estimate the peak power for N_s = 9 (PVs in series) and N_p = 250 (PVs in parallel) using $Pnpp = (N_S \times Vmpp) \times (N_P \times Impp)$
4. Select the DC-link voltage
5. Initialize the state, action, and reward for the RLA agent.
6. Initialize the duty cycle $Dint = 0.5$, $exb = -100$
7. Initialize the $0 < \alpha \leq 1$ fractional factor
8. Calculate the (t) and $\Delta e(t)$ based on the fractional value of α.
9. State-space X = $[V_{PV}, I_{PV}, P_{PV}, \Delta P_{PV}, \int \Delta P_{PV}, \alpha e(t), \Delta V_{DC}]$ + Action space U = [0,1]
10. Reward functions calculate using $[\Delta V_{DC}, \Delta P_{PV}, \alpha e(t), exb, \alpha \Delta e(t)]$
11. Provide these values to the network DQN

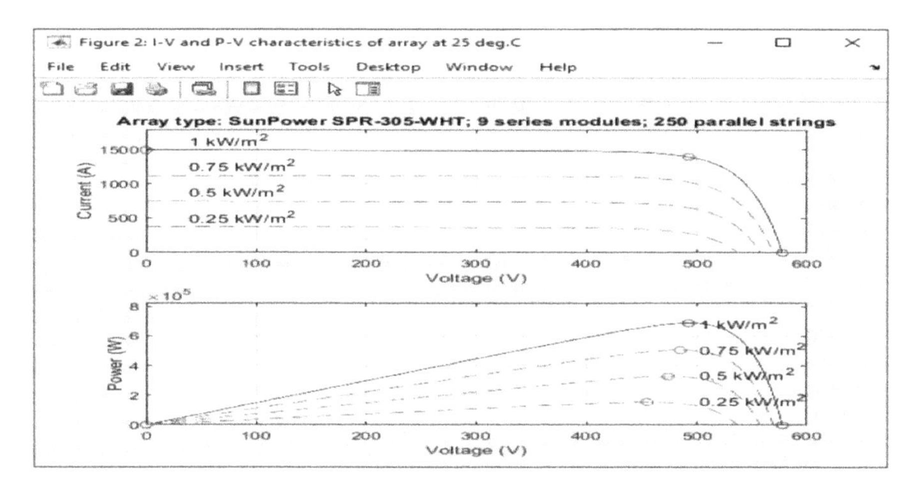

FIGURE 3.6 PV system under different irradiation condition.

12. Initialize/load Q and replay buffer R. β is the learning rate and γ is the discount factor.
13. for $j = 1$ *to* M do:
14. Get initial state $x0$
15. for t = 1 to T do:
16. Select action ut from the set defined
17. Execute the action ut
18. Get a new state $xt + 1$ and reward r
19. Store the transition $(xt, ut, ut + 1, ut + 1)$ in R
20. IF $|R| > N$
21. Update the network using Q-learning:
23. Update the network
24. **end if**
25. Set $xt = xt + 1$
26. **end for**

3.7 RESULT AND DISCUSSION

The main objective of fractional-order principle is to construct a novel model for tracking maximum performance. The design of MPP is done on MATLAB® (2019a). The initial duty cycle is $D_{int} = 0.5$. A Deep Q algorithm is used to train the RLA agent. The limits are set as (0,1). The action space for the RLA agent is designed in intervals over the step size of (0,0.01:1) resulting in a matrix size of 100 × 1. The learning rate is designed as 0.01with regularization L2 having values 0.0001 and a mini-batch size of N = 64. The implementation of reinforcement learning allows the model unlike parametric design variations to adapt the environmental impacts, and the training often takes place using a Deep Q learning algorithm, making it a model-free approach. This learning has the added benefit of being able to accurately monitor MPP after the network has learned. The efficiency of the cell is high, as the solar irradiance being as minimum as 250 W/m^2 produces a power of 0.2 kW/m^2 and it improves with the increase in sun radiations. The addition of the fractional order to the proposed construction aims to decrease monitoring time to the peak stage, maintain a steady output without oscillating around the MPP and reducing the THD output portion due to changes in the solar radiation, eventually affecting the efficiency of the photovoltaic device. The comparative analyses indicate that the model is successful in extracting maximum energy under different conditions of solar irradiation. The application of the proposed model to the actual data set means that this model can also be easily applied without a parametric adjustment in the model and generates full power even in severe winter conditions with the lowest solar irradiance. The designed model is RLA based; Figure 3.7 showcases the training of the DQN learning for the RLA model over an episode of 300 iterations. This training is done using the standard conditions of temperature and solar irradiance.

The model gets converged around 600 iterations. This illustrates the ability of the designed model to adapt quickly to the changes in environmental conditions. This completes the brief overview of the parameters of the model designed and the platform. The results section that follows includes two parts: the deigned model is compared against the various benchmark algorithm to showcase its robustness and

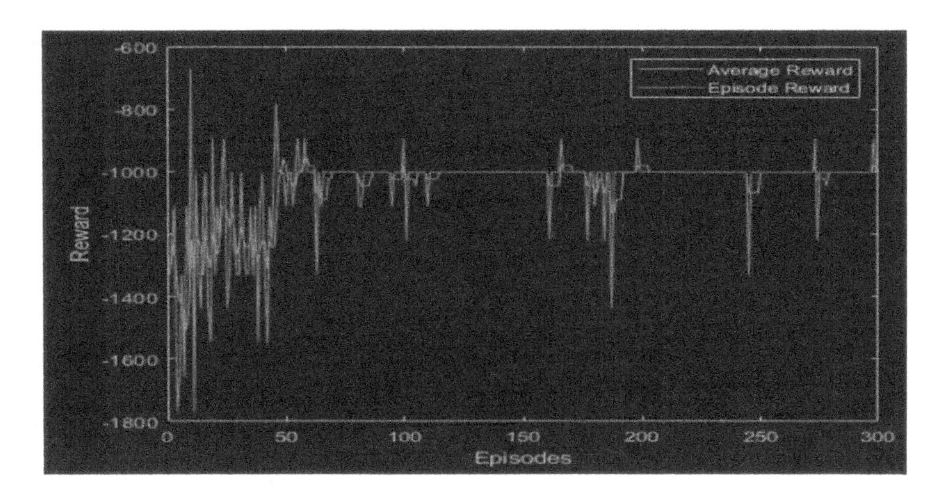

FIGURE 3.7 Proposed fractional-order DQN MPP control model training curve.

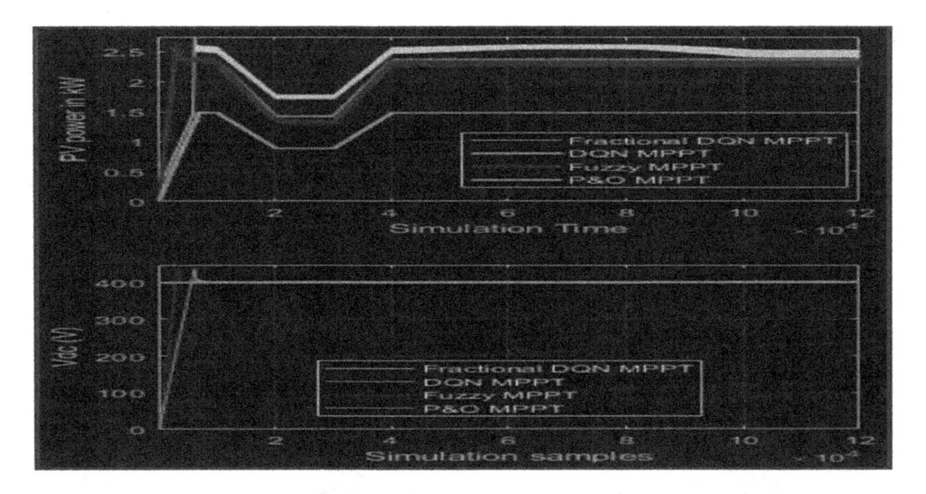

FIGURE 3.8 Comparative analysis of different algorithms for PV power and V_{DC}.

efficiency; and then the model is validated on a real data set to prove that, once trained, it is efficient to operate in different environmental conditions. Ding, Min et al. (2019) and Chu and Chen-Chi (2009) proposed a fractional-order method juxtaposed with other comparative benchmark work, especially trying to control the problem of MPP with RL. Santhiya et al. (2018) described the methods compared with MPP (P&O), fuzzy-MPP, and DQN-MPP. Here, with help of a controlling parameter fraction of change in the power is reverted into the circuit. This factor α is varied over the range as stated in the algorithm. P&O is one of the most widely used models for MPP control. The change in duty cycle and power along with voltage gets tested with every cycle of perturbation.

A reinitialization is done when violations occur. The comparison in terms of maximum power and voltage is shown in Figure 3.8. Total harmonic distortion (THD) is the amount of harmonic distortion present in any signal. It is defined as the ratio of the sum of all the harmonics present to the fundamental frequency. The presence of sunlight is a fluctuating occurrence that eventually leads to energy fluctuations undesirable for system performance. It is one of the serious concerns in the PV systems that are integrated between converter (DC/AC) and grid output systems as shown in Figure 3.9. The designed model with fractional control has been able to reduce THD components under the standard conditions. A comparison of the various methods based on THD has been shown in Table 3.5. This term is calculated here for all the comparative algorithms based on fundamental frequency and the first five harmonics encountered when a modified period-gram (Kaiser Window) is used. It can be observed that the proposed FODQN has the smallest THD component and also settles in just a few seconds. Also, the proposed algorithm is compared with other two designed algorithms in terms of various output characteristics at MPP under various solar irradiance conditions and keeping the temperature at a constant of 25°C.

Uniformity in the comparison values has been normalized using the standard characteristic of the PV array utilized by the designs. Through this analysis, we can conclude that the design outperforms in terms of maximum power in presence of distinct environmental conditions (see Table 3.6).

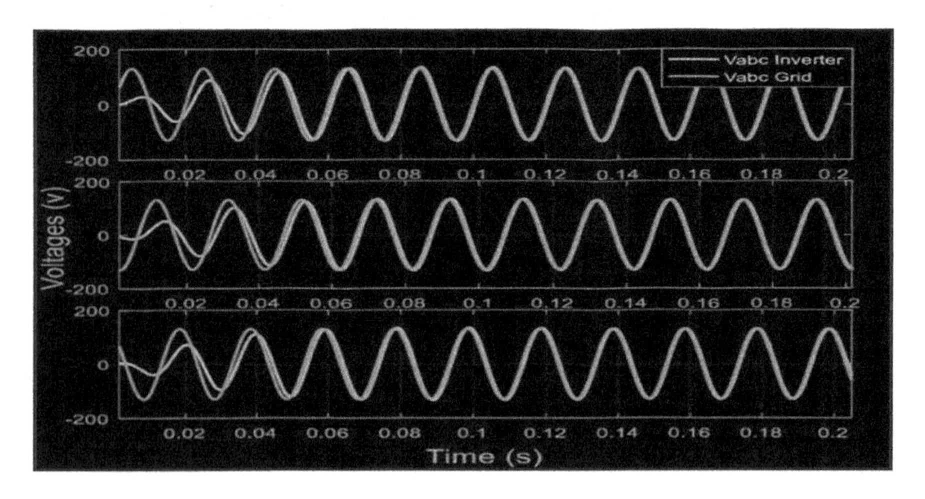

FIGURE 3.9 Line output voltage of the (DC/AC) converter and grid variable simulation results.

TABLE 3.5
Comparative Results of Various Methods Against THD and Settling Time

Parameters	Proposed Fractional DQN MPP	DQN MPP	Fuzzy MPP	P&O MPP
Power (Kw/m²)	2.71	2.7	2.25	1.6
THD	–15.49	–21.214	–6.732	2.24
Settling Time	0.122	0.241	0.1856	0.1856

TABLE 3.6
Comparative Results of Various Output Characteristics at MPP

MPP	Irradiations (W/m²)				
	250	500	750	1000	1500
SST [Ding et al. (2019)]					
Ppv	0.306	0.601	0.897	0.923	0.9855
Ipv	0.641	0.704	0.714	0.746	0.782
Vpv	0.194	0.422	0.640	0.841	0.940
FGFMG [Babaei et al. (2020)]					
Ppv	0.474	0.9384	0.9421	0.9546	0.9841
Ipv	0.6278	0.7005	0.6790	0.7661	0.8065
Vpv	0.1598	0.3517	0.5430	0.7742	0.8653
Proposed FODQN					
Ppv	0.2248	0.4341	0.6501	0.8645	1.07885
Ipv	0.3032	0.4595	0.5243	0.6997	0.8317
Vpv	0.2126	0.5187	0.7308	0.8465	0.8482

TABLE 3.7
Performance and Comparison of FODQN MPP Techniques with Other Fractional-Order Approaches

Work Proposed	Rise Time	Robustness	Deviation
Fixed FLC [Chu, Chen-Chi et al., (2009)]	0.21	Good	Good
Adaptive FLC [Eltamaly, Ali M (2015)]	0.156	Good	Good
ICM [Hukya Laxman et al., (2021)]	0.161	Bad	Bad
FO [Lin, Chia-Hung, et al., (2011)]	0.21	Bad	Bad
FOFLC [Yu,Kuo-Nan et al., (2015)]	0.031	Very Good	Very Good
FLC [Vanchinathan, K. (2012)]	0.154	Good	Bad
Proposed FODQN	0.223	Very Good	Good

Note: Bad, Good, and Very Good are the language labels created to compare the methods.

3.8 COMPARATIVE RESULTS OF VARIOUS OUTPUT CHARACTERISTICS AT MPP FOR DIFFERENT SOLAR IRRADIATION AT STANDARD TEMPERATURE CONDITIONS

In this section, we also compared the proposed work with other fuzzy methods on the basis of the rise time of voltage V_{DC}, robustness, and deviation. The method is compared to FOFLC, adaptive FLC, Fractional Order (FO), and IncCond Methods (ICM). The proposed method has a smooth transition, with a good tracking speed and steady-state stability. Also, a comparative settling time, a model-free approach, has produced good results under varying temperatures and irritation conditions.

These all point at the proposed method as a robust design. The deviation defined here is the difference between V_{DC} and the V_{ref}, thereby making it more vulnerable as compared to all the techniques existing in the literature as tabulated in Table 3.7. These values are taken for comparative analysis. These are based on simulations done by the respective authors using state-of-the-art schemes. They have adopted different PV models, so the linguistic approach is used to compare their work with our results.

3.9 CONCLUSION

The model for tracking maximum power is designed using reinforcement learning and the fractional-order concept. The introduction of reinforcement learning is done to make the model independent of parametric variations in the design to adapt to the environmental effects; the training also is done using a Deep Q-learning algorithm that makes the design a model-free approach. The added advantage of this learning is that once trained, the network can withstand any variation and still effectively track the MPP. The addition of fractional order in the proposed design aims to reduce the tracking time for reaching the peak level, maintain a steady-state output without oscillation around the MPP, and reduce the THD component in the output due to variations in solar radiation that eventually affect the PV array performance.

Testing of the design is carried out in two phases where it is compared with various benchmark existing algorithms and also tested using the real data set to capture the essence of the design under various environmental conditions. The lowest THD component and tracking time with maximum power output are achieved with FODQN compared to various algorithms. The comparative analysis confirms that the model is effective in harvesting maximum energy at different solar radiation conditions. The application of the proposed model on the real data set ensures that this model can be easily implemented in any area without any parametric change in the model and also produce maximum power even under the lowest solar radiance in extreme winter conditions where the presence of solar energy is lowest.

REFERENCES

Albert, J.R. and Stonier, A.A. (2020). Design and development of symmetrical super-lift DC–AC converter using firefly algorithm for solar-photovoltaic applications. *IET Circuits Devices Systems*, 14, 261–269. ISSN 1751-8598, doi:10.1049/iet-cds.2018.5292.

Albert, Johny Renoald and Vanaja, Dishore Shunmugham. (December 10 2020). Solar Energy Assessment in Various Regions of Indian Sub-continent [Online First], IntechOpen. doi:10.5772/intechopen.95118. https://www.Intechopen.com,online-first/solar-energy-assessment-in-various-regions-of-indian-sub-continent.

Babaei Milad, S., Yahyazadeh, Meisam, and Marj, Hassan Fatehi. (2020). Novel MPPT for linear-rotational sun-tracking system using fractional fuzzy grey-based sliding mode control. *Iranian Journal of Science and Technology, Transactions of Electrical Engineering*, 44, 1–23.

Behera, Manoja Kumar and Saikia, Lalit Chandra. (2020). A new combined extreme learning machine variable steepest gradient ascent MPPT for PV system based on optimized PI-FOI cascade controller under uniform and partial shading conditions. *Sustainable Energy Technologies and Assessments*, 42, 100859. doi:10.1016/j.seta.2020.100859. https://www.sciencedirect.com/science/article/pii/S2213138820312868.

Benlahbib, Boualam, Bouarroudj, Noureddine, Mekhilef, Saad, Abdelkrim, Tameur, Lakhdari, Abdelkader, and Bouchafaa, Farid. (2018). A fuzzy logic controller based on maximum power point tracking algorithm for partially shaded PV array-experimental validation. *Elektronika ir Elektrotechnika*, 24, 4. ISSN 1392-1215, doi:10.5755/j01.eie.24.4.21476.

Bubalo, M., Bašic, M., Vukadinovic, D., and Grgic, I. (2021). Experimental investigation of a standalone wind energy system with a battery-assisted Quasi-Z-SourceInverter. *Energies*, 14, 1665. doi:10.3390/en14061665.

Change, I.P.O.C. (2014). IPCC. *Climate change*. https://www.mathworks.com/help/matlab/ref/.

Chu, Chen-Chi and Chen, Chieh-Li. (2009). Robust maximum power point tracking method for photovoltaic cells: A sliding mode control approach. *Solar Energy*, 83, 8, 1370–1378.

Dhivya, M. and Renoald, A.J. (2017). Fuzzy grammar based hybrid split-capacitors and split inductors applied in positive output Luo-converters. *International Journal of Scientific Research in Science, Engineering and Technology*, 3, 327–332. http://ijsrset.com/IJSRSET173174.

Ding, Min, Lv, Dong, Yang, Chen, Li, Shi, Fang, Qi, Yang, Bo, and Zhang, Xiaoshun. (2019). Global maximum power point tracking of PV systems under partial shading condition: A transfer reinforcement learning approach. *Applied Sciences*, 9, 13, 2769. doi:10.3390/app9132769.

Eltamaly Ali, M. (2015). Performance of smart maximum power point tracker under partial shading conditions of photovoltaic systems. *Journal of Renewable and Sustainable Energy*, 7, 4, 043141.

Gautam, Abhishek and Saini, R.P. (2020). A review on technical, applications and economic aspect of packed bed solar thermal energy storage system. *Journal of Energy Storage*, 27, 101046. ISSN 2352-152X, doi:10.1016/j.est.2019.101046. https://www.sciencedirect.com/science/article/pii/S2352152X19309533.

Gordon, Jeffrey M., Fasquelle, Thomas, Nadal, Elie, and Vossier, Alexis. (2021). Providing large-scale electricity demand with photovoltaics and molten-salt storage. *Renewable and Sustainable Energy Reviews*, 135, 110261. ISSN 1364-0321, doi:10.1016/j.rser.2020.110261. https://www.sciencedirect.com/science/article/pii/S1364032120305505.

Goswami, Y. and Sharma, S. (2021). Graphene Reinforced Biopolymer Nano-composites in Energy Storage Applications. In: Sharma, B. and Jain, P. (eds.) *Graphene Based Biopolymer Nano-Composites*. Composites Science and Technology. Springer, Singapore. doi:10.1007/978-981-15-9180-8_13.

Hong, Ying-Yi, Beltran, Angelo A., and Paglinawan, Arnold C. (2018). A robust design of maximum power point tracking using Taguchi method for stand-alone PV system. *Applied Energy*, 211, 50–63. ISSN 0306-2619, doi:10.1016/j.apenergy.2017.11.041. https://www.sciencedirect.com/science/article/pii/S0306261917316161.

Khan, Saadullah, Ahmad, Aqueel, Ahmad, Furkan, Shemami, Mahdi Shafaati, Alam, Mohammad Saad, and Khateeb, Siddiq. (2018). A comprehensive review on solar powered electric vehicle charging system. *Smart Science*, 6, 1, 54–79. doi:10.1080/23080477.2017.1419054.

Laxman, Hukya, Annamraju, Anil, and Srikanth, Nandiraju Venkata. (2021). A grey wolf optimized fuzzy logic based MPPT for shaded solar photovoltaic systems in micro-grids. *International Journal of Hydrogen Energy*, 46, 18, 10653–10665. doi:10.1016/j.ijhydene.2020.12.158. https://www.sciencedirect.com/science/article/pii/S036031992034790X.

Lin, Chia-Hung, Huang, Cong-Hui, Du, Yi-Chun, and Chen, Jian-Liung. (2011). Maximum photovoltaic power tracking for the PV array using the fractional order incremental conductance method. *Applied Energy*, 88, (12), 4840–4847.

Mazzeo, Domenico, Baglivo, Cristina, Matera, Nicoletta, Congedo, Paolo M., and Oliveti, Giuseppe. (2020). A novel energy-economic-environmental multi-criteria decision-making in the optimization of a hybrid renewable system. *Sustainable Cities and Society*, 52, 101780. ISSN: 2210-6707, doi:10.1016/j.scs.2019.101780. https://www.sciencedirect.com/science/article/pii/S221067071931011X.

Merenda, Massimo, Iero, Demetrio, Carotenuto, Riccardo, Corte, Della, and Francesco, G. (2019). Simple and low-cost photovoltaic module emulator. *Electronics*, 8, 12, 1445. doi:10.3390/electronics8121445.

Minai, A.F. and Malik, H. (2021). Metaheuristics Paradigms for Renewable Energy Systems: Advances in Optimization Algorithms. In: Malik, H., Iqbal, A., Joshi, P., Agrawal, S., and Bakhsh, F.I. (eds.) *Metaheuristic and Evolutionary Computation: Algorithms and Applications*. Studies in Computational Intelligence, vol 916. Springer, Singapore. doi:10.1007/978-981-15-7571-6_2.

Nižetić, Sandro, Djilali, Nedjib, Papadopoulos, Agis, and Rodrigues, Joel J.P.C. (2019). Smart technologies for promotion of energy efficiency, utilization of sustainable resources and waste management. *Journal of Cleaner Production*, 231, 565–591. ISSN 0959-6526, doi:10.1016/j.jclepro.2019.04.397. https://www.Sciencedirect.com/science/article/pii/S0959652619314982.

Nkambule, M.S., Hasan, A.N., Ali, A. et al. (2021). Comprehensive evaluation of machine learning MPPT algorithms for a PV system under different weather conditions. *Journal of Electrical Engineering & Technology*, 16, 411–427. doi:10.1007/s42835-020-00598-0.

Pakkiraiah, B. and Sukumar, G. Durga. (2016). Research survey on various MPPT performance issues to improve the solar PV system efficiency. *Hindawi Publishing Corporation Journal of Solar Energy*, 2016, Article ID 8012432, 20 pages.

Rezkallah, M., Hamadi, A., Chandra, A., and Singh, B. (October 2015). Real-time HIL implementation of sliding mode control for standalone system based on PV array without using dumpload. *IEEE Transactions on Sustainable Energy*, 6, 4, 1389–1398. doi:10.1109/TSTE.2015.2436333.

Roy, Priyanka, Sinha, Numeshwar Kumar, Tiwari, Sanjay, and Khare, Ayush. (2020). A review on perovskite solar cells: Evolution of architecture, fabrication techniques, commercialization issues and status. *Solar Energy*, 198, 665–688. ISSN 0038-092X, doi:10.1016/j.solener.2020.01.080. https://www.sciencedirect.com/science/article/pii/S0038092X20300888.

Santhiya, K., Devimuppudathi, K., Santhosh Kumar, D., and Albert, J.R. (2018). Real time speed control of three phase induction motor by using lab view with fuzzy logic. *Journal on Science Engineering and Technology*, 5, (2), 21–27. http://jset.sasapublications.com/wp-content/uploads/2018/04/1841019.pdf.

Shunmugham Vanaja, D., Albert, J.R., and Stonier, A.A. (2021). An experimental investigation on solar PV fed modular STATCOM in WECS using intelligent controller. *International Transactions on Electrical Energy Systems*, 2021, e12845. doi:10.1002/2050-7038.12845.

Thenmalar, K, Albert, J.R., Begam, Muhamadha, and Madhumathi, P. (2021). Power quality improvement in modular multilevel inverter using for different multicarrier PWM. *European Journal of Electrical Engineering and Computer Science*, 05, 02, 2021. ISSN: 2736-5751, doi:10.24018/ejece.2021.5.2.315.

Vanchinathan, K. (2012). Novel voltage stability analysis of a grid connected–photovoltic system. *Journal of Global Research in Computer Science*, 3, (7), 38–43.

Yu, Kuo-Nan, Yau, Her-Terng, and Liao, Chi-Kang. (2015). Development of a fractional-order chaos synchronization dynamic error detector for maximum power point tracking of photovoltaic power systems. *Applied Sciences*, 5, (4), 1117–1133.

Zhou, Yuekuan, Cao, Sunliang, Jan Hensen, L.M., and Lund, Peter D. (2019). Energy integration and interaction between buildings and vehicles: A state-of-the-art review. *Renewable and Sustainable Energy Reviews*, 114, 109337. ISSN 1364-0321, doi:10.1016/j.rser.2019.109337. https://www.sciencedirect.com/science/article/pii/S1364032119305453.

4 Challenges and Opportunities for Predictive Maintenance of Solar Plants

K. P. Suresh, R. Senthilkumar and S. Saravanan
Sri Krishna College of Technology, Coimbatore, India

M. Suresh
Kongu Engineering College, Erode, India

P. Jamuna
Nandha Engineering College, Erode, India

CONTENTS

DOI: 10.1201/9781003202288-4

LEARNING OUTCOME

i. To study the importance of solar power plant periodic maintenance along with dependencies and its challenges in power generation
ii. To know about the classification of maintenance and its protection of solar PV plants
iii. To study emerging techniques used in predictive maintenance of solar plants

4.1 INTRODUCTION

Nowadays, solar power generation appears to be the most optimal alternative to fossil fuels to meet power demand worldwide, mainly due to its being plentiful and easily accessible, even with some fluctuations throughout the year. Another advantage of providing access to electricity via solar energy is that it can be done in the regions where grid-connected electricity is not readily or at all available. Worldwide today, constructed and operating solar power plants provide about 150GW of electricity annually [1]. Effectiveness of a solar plant depends on proper maintenance, and a normal lifetime of a solar power plant is between 25 and 40 years when effectively maintained. The main objectives of the predictive maintenance of a photovoltaic (PV) system are to decrease maintenance costs and increase system efficiency and power storage capacity. This chapter discusses challenges and opportunities in predictive maintenance of a solar photovoltaic system.

There are different methods available in operation and maintenance based on nature of the installation of a PV system. The main factors considered in the operation and maintenance include:

1. PV system performance metrics
2. Standard definitions and formula
3. Differences in plant location and weather conditions

Due to the current state of solar power generation technology, most of the investment in the industry is geared toward minimizing the risks of faults and downtime than increasing production. Standardization of operation and maintenance is the most optimal way to achieve shorter downtimes and lower costs associated with facilitating repairs and bringing the plant back online [1].

Technical guidelines for operation and maintenance are generally provided by different organization in cooperation with business-industry groups, such as

- Institute of Electrical and Electronics Engineers (IEEE)
- American Society for Testing and Materials (ASTM)
- International Electrotechnical Commission (IEC)

Representatives from IEEE, IEC, and ASTM frequently collaborate to provide best practices for operation and maintenance in renewable energy systems by way of creating standard certification requirements and recommendations.

An integrated PV system with battery storage is associated with higher costs and more complex day-to-day operations when compared to a grid-integrated PV system. However, given that one of the main advantages of solar power generation is more effective power generation in regions not connected to a conventional power grid, many researchers concentrate their work on finding ways to reduce the complexity operation and maintenance of battery-storage PV systems. Figure 4.1 shows the project development status in a PV system.

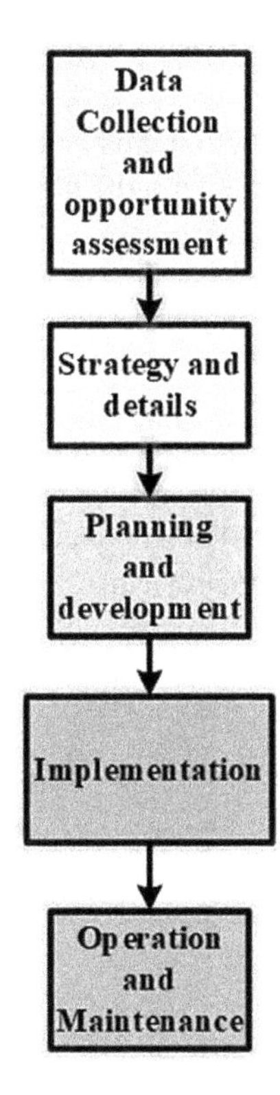

FIGURE 4.1 PV system project development status.

4.2 SOLAR PLANT OPERATION

A solar plant consists of good-quality PV panels, converter, inverter with connecting wires, and utility meters combined with a monitoring system. Moreover, the predictive maintenance of solar plants is important to enhance the lifetime and efficiency of a plant. Lack of awareness about PV plant maintenance leads to less efficiency. The importance of predictive maintenance includes regular monitoring, performance analysis compared with with previously collected historical data, and prevention of possible failures and energy loss in generation. The advantages of predictive maintenance of solar plants include safety during the entire life span of the plant, reduction in repair and (consequently) idle time, and savings on maintenance and spare parts. In predictive maintenance, regular inspection and remote monitoring systems with sensors are enabled based on recommendations from equipment manufacturers. The main challenges for predictive maintenance of solar plants are identifying immediate changes in performance behavior and remote-control system to control significant parameters such as active power control, reactive power control, frequency control, and voltage control. In solar panel maintenance, the factors to be considered include:

1. Effective performance
2. Maintenance based on maximizing power delivery
3. Scheduling maintenance
4. Assessment of performance and trends
5. Operating the system interface with the power grid
6. Security monitoring

A PV system requires the following maintenance procedures:

- Administration of both operation and maintenance, which includes effective implementation, quality control, and feedback system
- Frequency of preventive maintenance and its scheduling
- Replacement and repair of failed components
- Condition-based corrective maintenance

Operation and maintenance dependencies in PV systems are:

1. Electrical system
2. Inverter system
3. Rooftop system
4. Rack system
5. Ground mounting system
6. Tracking mount system
7. Environment conditions

4.2.1 Electrical System

In a PV power generating system both DC and AC components play an important role in determining system capacity and reliability. Wiring management is an important

FIGURE 4.2 Maintenance of electrical wiring in a PV system.

factor in maintenance and prevention of physical damage. In a PV system, the connections available in one module-to-module and then module-to-control unit box in open air. This will lead to damage due to external environmental effects. In this wiring system, problems can be easily detected and solutions identified. Dealing with conducting wires buried in the ground is more expensive and time- and effort-consuming in terms of finding faults. In a PV system, the electrical cables, insulated wire ties, bushings, and wire rack parts are exposed to direct sunlight, and because of this, adequate inspections are required to assess possible damage due to wind, temperature changes, and other external effects. In a maximum power point tracking (MPPT)-based PV system, which has continuous movement in rack to extract maximum power, that movement may cause deformation and other damage to mounted parts carrying electrical wires, creating the need for repair. Figure 4.2 shows the wiring without any rack system in a PV power generation system. Frequent maintenance will be required to prevent and fix damage caused by animal life, dust, dirt, and excessive humidity.

4.2.2 Inverter System

There are different types of inverters installed in a PV power generation system. They include:

- String inverter
- DC optimizer/inverter
- Central inverter
- Microinverter

Inverter is an important component in PV system connecting to the AC load, which mostly depends on the different types of inverter topology. The PV system performance can be reduced due to inverter failure. One of the most common reasons for inverter failure is a clogged filter, which often occurs in areas of lush vegetation. Spare parts for inverters are expensive, and when considered together with time and labor costs, it becomes more cost effective to replace the entire inverter rather than attempting to fix an existing malfunctioning one. The production equipment must also meet the standards of the National Fire Protection Agency (NFPA) during the maintenance period. The inverter topology selection is important based on PV system capacity and also for replacement if any damage to the inverter occurs.

4.2.3 Rooftop System

Modern solar power technology makes a rooftop solar system into a dual-purpose installation, both to protect the building as part of the roof and to generate electricity. Roof maintenance related to a PV system is required to maintain roof warranty. The roofing company must provide the required specifications that must be followed precisely to rectify problems in a PV system. Periodic maintenance should be needed for all materials, insulation compression, frame, polyvinyl chloride (PVC) and thermoplastic polyolefin (TPO) in a rooftop system. There are different types of roofing available in a PV system with service costs based on roof size, site location, accessibility, and labor cost. The detailed service costs with corresponding roof types are presented in Table 4.1.

The rooftop maintenance system includes roof warranty along with agreement provided by the roofing company, ice dams removal, and debris removal. If these snow, ice, and debris are not properly removed, the annual generation is reduced 5% to 15%, not to mention the potential damage to the rooftop itself and, consequently, to a PV panel.

TABLE 4.1
Service Cost with Different Types of Roof (Home Advisor 2018)

S. No	Different Types of Roof	Repair Materials ($\$/m^2$)	Repair Labor (h/m^2)
1.	Built-up, Bituminous	15	1.5
2.	Asphalt Shingle	15	1.0
3.	Thermoplastic Polyolefin (TPO)	20	1.0
4.	Ethylene Propylene Dlene Monomer (EPDM)	20	1.0
5.	Polyvinyl Chloride (PVC)	20	1.0
6.	Styrene-Butadiene-Styrene (SBS)	20	1.0
7.	Composite Shingle	25	1.0
8.	Wood Shingle	40	2.0
9.	Slate	50	1.0
10.	Metal Roof	50	0.5
11.	Tile	50	1.0

4.2.4 Rack System

PV panels are mounted on a supporting frame called roof deck or roof rack system, which usually is tilted 10 to 20 degrees to withstand wind loads. The panel tilt direction should be either south facing or north facing, which delivers more power per square foot. The maintenance of the rack system includes roofing material, compressible insulation, roof membrane, roof deflection, roof drainage, and interconnections of roof elements. An automatic tilt rack system requires more maintenance than a stationary rack system [2]. An MPPT-based PV system has an automatic rack tilt with respect to sun radiation, which needs periodic predictive maintenance.

4.2.5 Ground-Mounted System

In ground-mounted systems, the cleaning requirements, electric wiring patterns, and maintenance of drainage are included in the design. The ground maintenance issues include grass mowing, snow removal, cleanup of dust and debris, drainage of fluids such as rainwater and pesticide sprays, and axis tracking. Extra care must be taken with any kind of mechanical cleaning, such as snow, ice, or debris removal, so as not to damage the PV panels. Figure 4.3 shows that the drainage outlets should be properly planned before PV panel installation, otherwise the construction of a proper drainage becomes a very difficult and expensive task.

FIGURE 4.3 Ground-mounted PV system.

4.2.6 Tracking Mount System

In PV systems, the tracking mount plays an important role in determining the performance and improving outcome expectations. The mount also requires periodic maintenance due to interconnection of moving parts associated with sensors and actuators. An automated tracking mount increases the overall cause of the system, but it proves to be cost efficient in terms of improvements in power generation, compared to a nonautomated tracking system, by allowing immediate adjustments to the panel tilt as the sun moves across the sky throughout the day. There are different types of tracking systems available, including

- Single axis
- Dual axis
- Centralized tracking
- Decentralized tracking

Single-axis tracking means the system allows rotation along only one directional axis. A dual-axis mount is able to rotate along two different axes, usually perpendicular to each other. Centralized and decentralized tracking mechanisms have their own advantages in terms of electrical connections, automatic controls, sensors and actuators, and monitoring for preventive maintenance [3].

4.2.7 Environmental Conditions

Operation and maintenance of PV systems are affected by different environmental conditions. The effect of such conditions can be minimized by careful construction area selection so as to minimize the risks of contamination and/or damage [4]. Some environmental effects include

- High (or low) temperature
- Heavy wind
- Salty, exceedingly humid air
- Heavy snowfall
- Heavy rainfall
- Humidity
- Pollen
- Bird populations/other wildlife
- Sand/dust
- Diesel soot
- Industrial emissions

Periodic cleaning with water and manufacturer-recommended mild cleaning solution should performed regularly; care should be taken not to damage the PV array during cleanup. Figure 4.4 shows soil dust formation on a PV panel. Nowadays, robotic systems are available for cleaning large PV plants with sensors. In areas of bountiful avian life, panel pollution can be reduced by installing bird spikes to prevent roosting. Specialized cleaning may be required to check for and remove bird nests during the nesting season.

FIGURE 4.4 Soil dust on PV panel.

4.3 SOLAR PLANT MAINTENANCE

4.3.1 PREDICTIVE MAINTENANCE

With the increased proliferation of PV power plants, spot maintenance becomes less and less efficient and cost effective due to exorbitant manpower requirements. Consequently, maintenance approaches that do not require manual diagnostic are becoming more popular, including FMEA, machine learning, and real-time sensors [5]. The recent approaches for this predictive maintenance of a PV plant include three methods. First is direct visual inspection of all components, I-V characteristics analysis of the entire plant, infrared thermography, and comparison current output with actual generation capacity of the plant. Second is machine learning and artificial intelligence–based forecasting methods, which offer moderate levels of effectiveness in terms of fault detection. Third is smart remote monitoring and control unit with wireless sensor networks. Out of these three methods, the wireless sensor method is most expensive, but it also offers the best results in terms of predictive detection [6]. Figure 4.5 shows the current approaches available and predictive maintenance for PV system.

4.3.2 Classification of Maintenance

Maintenance is a "combination of all technological, managerial, and procedural activities taken during a product's lifespan in order to keep it in, or restored it to, a condition, where it can carry out specific function" [7]. The maintenance plan explains how these activities are carried out and integrated to meet the maintenance goals. The main objectives of maintenance are increasing supply, lowering overall costs, improving product efficiency, preserving the environment, and improving protection [8]. Maintenance can be preventive or corrective, as shown in Figure 4.6.

4.3.2.1 Preventive Maintenance

Preventive maintenance is performed to decrease the probability of an item failure or loss of functionality. Condition-based maintenance and predetermined maintenance and are two types of preventive maintenance. It can be scheduled at fixed intervals so as to minimize interruption of normal operations, or it can be based on another consumption parameter. When degradation measures reach a predefined threshold, preventive condition-based maintenance is used. The ability to track component state is a requirement of condition-based maintenance. A thorough understanding of the failure characteristics of the elements is needed in order to predict failures.

4.3.2.2 Corrective Maintenance

Once a fault has been identified, corrective maintenance is performed. Corrective measures need to be taken as promptly as allowed by regulations. Periodically

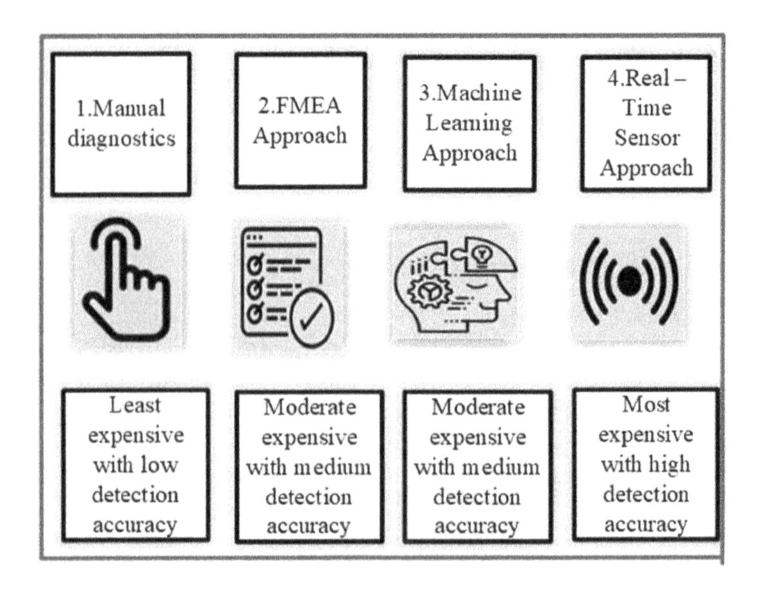

FIGURE 4.5 Opportunities for predictive maintenance.

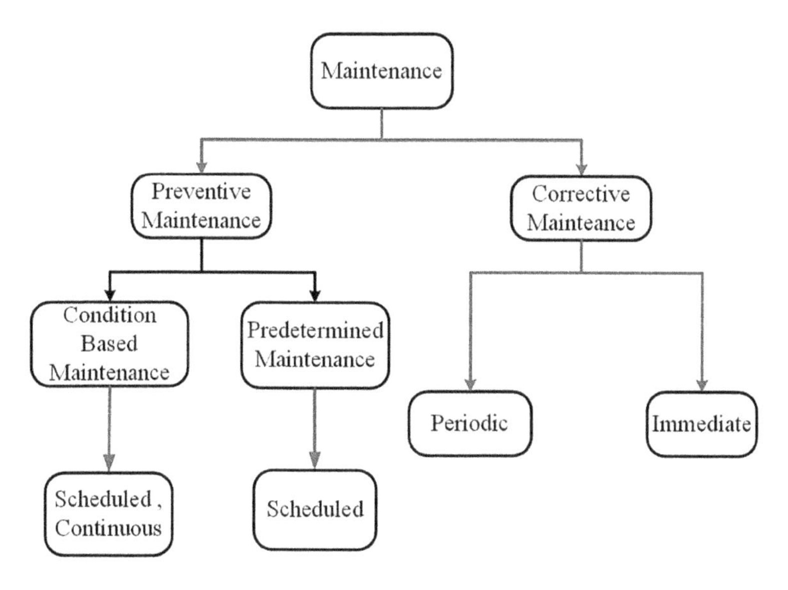

FIGURE 4.6 Classification of maintenance.

scheduled repair is a form of delayed corrective maintenance aimed to reduce costs and downtime by "bundling" multiple smaller repairs into one regularly scheduled service call.

4.4 METHODS OF PREVENTIVE MAINTENANCE

In pre-failure mode, preventive maintenance is used to reduce larger losses that can result after from damage that affects the stochastic potential of operations [9]. The four schemes used for this method of maintenance are summarized in Figure 4.7.

4.4.1 PV Panel Cooling

Individual cells within a PV panel are often exposed to different levels of sunlight intensity, resulting in temperature variations across the panel. This should not affect the actual operation of the device, but if temperature differences continue to increase, the form factor and the open-circuit voltage will decrease. The reciprocal relationship between temperature gaps as well as chemical bonding within each cell cause this phenomenon. There are two cooling mechanisms available to decrease the temperature prior to device failure.

The ground-coupled centralized panel cooling system circulates cold air through a fan and directs it onto the PV panels' substrate, causing the dissipation of heat. A heat sink achieves the same effect by filtering away excess stored heat from a certain area of the PV series. Continuously cooling a PV panel, regardless of the process, means extending its life cycle. Since semiconductive materials that allow electrons to pass through as light strikes them do not tolerate heat, cooling provides a refresher for these components to perform at their best. Among the various cooling methods, immersing PV panels in a dielectric medium or directly spraying cold water on their

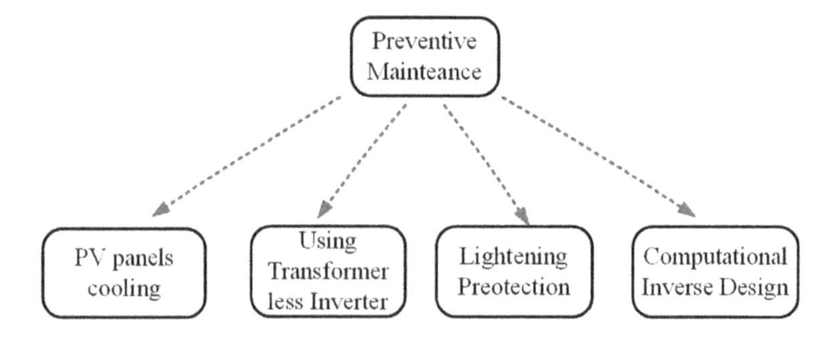

FIGURE 4.7 Schemes of preventive maintenance.

surfaces is the most preferred, since it combines the benefits of washing and cooling the panels.

4.4.2 Transformer-Less Inverter

The major power used at the end of any PV grid is usually in AC form, but coming directly from a PV array, it is usually in DC form. Inverters convert DC current to AC current. Transformers have been used in inverter models to obtain galvanic separation between AC and DC modules, and they are often used to increase DC voltage from PV output power to greater levels. This factor decreases the overall inverter performance, even though the machine will operate perfectly without any failures. To put it another way, when transformers are required, we can switch to transformer-less inverters, which attains the galvanic separation between DC and AC parts. This method is more concerned with the electrical efficiency of a PV system than with cell activity, as it is with cooling systems. Using a transformer-less inverter can rectify transformer-related issues, resulting in a more structured power supply. In PV designs, a transformer-less inverter is preferred despite the lack of galvanic separation.

4.4.3 Lightning Protection

While a PV system can function optimally without certain preventive measures, their development facilitates the prevention of any destruction from rare but catastrophic weather events, such as lightning strikes. As lightning strikes, it causes immediate over-voltage and thermal burn to PV modules, deteriorating the system's electric features and dropping the output voltage to zero. As a result, the effective lightning safety design ensures uninterrupted performance of the system.

The protective mechanism can be separated into two parts, internal and external, as seen in Figure 4.8. In the external lightning protection, a small impedance metal rod is positioned in the ground, which creates a fast way for any surge to flow in instead of disrupting devices when that massive amount of energy passes through the components of the system. In the internal structure, the separation interval is achieved by leaving sufficient space between the cells' ionization paths so that they are not

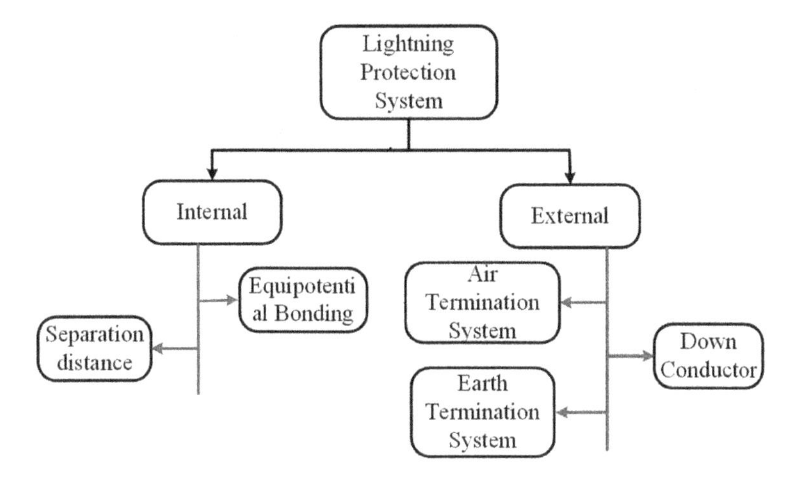

FIGURE 4.8 Lightning protection mechanism.

fully combined when a strike occurs. Lightning safety is recommended only for the areas of frequently occurring thunderstorms, similar to a snow-clearing strategy being part of a proactive maintenance system only in areas of regular heavy snowfall during the cold season.

4.4.4 Computational Inverse Design

To eliminate the ineffective trial-and-error method of determining the most feasible solutions for a specific input, a material characterization technique that improves the relationship between structural and electro-optical properties may be used. The transformation from a microscopic level to a desired chemical composition is possible due to numerical simulations. This is used to find a unique setup that is most similar to the desired functionality. When measuring a small proportion of the million-level potential configurations, this approach chooses the model that has the specific target features.

A commonly produced PV panel will function without such a design, but it introduces a potential failure prevention tool by selecting cell building blocks, then calculating the target property, and eventually using an optimization technique to achieve the desired level of the performance of the system. This is to make flexible cells that can withstand harsh environments in order to avoid potential failures and maximize the fabrication phase. For context, a properly identified triple-junction cell feature enables the absorption of various wavelengths of light. As a result, under different shading conditions, each PV array structure will produce more centralized power at the cell level.

A properly identified triple-junction cell feature enables the absorption of various wavelengths of light. As a result, under different shading conditions, each PV array structure will produce more power at the cell level. The designed chemical architecture for PV cells may aid in preventing improper nonlinear activity in them during future operations. The computational inverse model is aimed at PV manufacturers taking into consideration the existing techniques for producing stable PV cells that can withstand high temperatures and humidity.

The most usable strategy is preventive maintenance because it reduces the risk of faults and, consequently, the need for all other strategies. By selecting reliable PV cells, a transformer-less inverter, and other factors, a good initial PV design process can be added to preventive maintenance. It is the most cost-effective strategy compared to the others available, since its model requirements are part of the PV system's initial installation cost. When compared to corrective action after a fault occurs, preventive maintenance represents a regular, steady, predictable cost, unlike multiple instances of corrective maintenance.

4.4.5 Machine Learning and Artificial Intelligence–Based Forecasting Methods

This process is usually quite expensive and provides only average detection precision. It is based on estimating the solar energy by localized forecast for a short duration using microclimate elements and regressive process, as well as identifying parametric variables with higher impact rates or launch time-based weather data. Other analysis tools include Artificial Neural Network (ANN), Autoregressive Integrated Moving Average (ARIMA), and other methods to raise prediction accuracy by considering the time of day, sky cover, atmospheric pressure, and wind direction. Using ANN models to forecast the PV energy generation can also explore the relationship between the PV energy impact and weather station elements. Analyzing the effects of data from weather stations and comparing variable radiance levels with NWP predictions shows the importance of localization on solar energy forecasts.

It is essential to assess the weather elements with absolute accuracy to make the exact solar energy predictions [10]. Many approaches have been devised by researchers to forecast the weather elements that influence solar energy production. Multiple simulation methods for weather data have been developed as a result of the estimation of weather elements and solar energy. Various elements that disturb the amount of solar energy being produced are considered during the solar energy production and the factors affecting it. The weather elements include, but are not limited to, cloud cover, wind speed, temperature, and radiance levels. Many simulation and estimation models have been devised to detect any patterns in weather data that can aid in the prediction of these weather elements in advance, which can also be used to measure the amount of solar energy produced.

The amount of incoming solar radiation has a significant impact on the average efficiency of solar panels [11]. As a result, the capability of a solar array to collect sunlight is the most important factor impacting its performance and efficiency of any PV array. If there is no sunlight, no electricity can be produced by solar energy technologies. Solar panel module temperature and ambient temperature are two other important factors that affect the efficiency of solar energy systems.

Various models have been devised that incorporate past data and statistical weather forecast outlines to estimate temperature, specifically environmental temperature in the case of solar energy estimation. In [12], the authors used neural networks to build an atmospheric temperature model. This model produced a self-organizing feature map (SOFM) that was used to create multilayer perception (MLP) networks, predictably called SOFM-MLP. According to this study, in order to forecast atmospheric parameters more effectively, the choice of features used in this model is critical.

In [13], the authors proposed the multimodel super-ensemble forecast center to be used to improve weather and seasonal predictions. Lazos et al. [14] devised a hybrid weather prediction model made from a combination of statistical and numerical short-term estimation systems. In [15], the authors suggested that the autoregressive modification method is a suitable model that can differentiate the estimated values and actual values for the future temperature predictions. Thus, the research mentioned in this section is improving the predicting weather elements used to enhance solar energy production forecasting.

4.4.6 Smart Remote Monitoring of PV Plant

Various commercial systems are available to track both small and large grid-connected PV installations. Many firms, such as Fat Spaniel Technologies, SMA Solar Technologies, Access Networks, Morningstar Corporation, Fronius International GmbH, or SolarMax, sell commercial management applications for PV devices. Their implementation is typically provided on Software as a Service (SaaS) basis, which is a web-based service that provides the necessary features [16]. End-users can access transient phenomena as well as historical data on PV performance using this web application. Furthermore, some specialized systems have more useful features such as warnings or alerts when breakdowns or problems occur. Commercial options, on the other hand, have certain disadvantages [17]. Devices that measure and capture data use an excessive amount of resources and have large storage requirements. Commercial software does not fulfill all the necessary implementation criteria, so additional functionalities, such as customized performance evaluations or demand projections, cannot be applied from the user end.

4.4.6.1 Linux RTAI Monitoring System

Chaindone implemented a Linux Real Time Application Interface (RTAI)-based platform for detecting, recording, and transmitting weather and electrical data from a PV plant [16]. Electrical measurements, weather information, performance of the PV device, and performance of the inverter over different operating phenomena were among the data collated from the inverter and PV array via proper sensors. Working with such a machine requires a deeper understanding of MATLAB® and Simulink, as well as Linux for signal computation functions and other activities.

4.4.6.2 ZigBee-Based Monitoring System

In [17], the author suggested using ZigBee for wireless monitoring and control of Photovoltaic Distributed Generator for microgrid. If a transceiver is removed, the ZigBee module will establish a network of its own. Full Function Device (FFD) and Reduced Function Device (RFD) are the two components of the ZigBee scheme. FFD is the key device feature for communicating with other modules and for setting up addressing plans and keeping position tables. FFD can send data, locate nearby gadgets, and create pathways to various hubs. Only a nearby network is connected to the RFD. When the module must receive or send data, it usually wakes up. When buying a tracking system, device accuracy is a critical consideration for most consumers. On the other hand, ZigBee technology is unstable in terms of security. Mid-range hackers can be able to get into it. In addition, ZigBee's RF features have a range restriction. An obstruction can cause RF transmission to be disrupted.

4.4.6.3 NI-Based Monitoring System

Data Acquisition Systems (DAQ) have the capability to simultaneously measure and store data from several sources. There are anywhere between 8 and 32 peripheral channels in the system. Each channel has its own ADC. Data is collected in parallel, and events in each channel can be computed in real time. National Instruments (NI) products are based on real-time testing to assess the efficiency of PV systems. In response to a market inquiry, NI has developed a tracking system for PV modules/cells that is both cost effective and capable of assisting in the more accurate testing of modules [18].

4.4.6.4 ARM Processor-Based Monitoring System

Traditional costly monitoring systems adhere to IPC requirements for PV monitoring, and as a result of failure incidence and connectivity interruption, maintenance and replacement of the entire setup are difficult. In [19], the author recommended the LM3S9B92 (ARM Cotex-M3 processor) as a controller interface for grid-connected PV system data monitoring using LwIP. The PV system's data can be logged and processed, and faults can be recognized by the system. The monitoring system was created with LwIP and Cortex-M3 and included the following features: (1) controller modules for optimization; (2) standard Ethernet for high-speed data transfer; and (3) an easy-to-use user interface. Inverter, remote PC, and other part connections make up the system. The controller passes data to the network controller module after collecting data from different sensors within inverters. The information is then prepared and posted.

4.4.6.5 Satellite-Based Monitoring System

In [20], the author identified a monitoring device that sends data to the web through satellite ARGOS. The ARGOS satellite sends data at a rate of 10 bytes every 15 minutes [20]. The permeability circle of the satellite is around 5,000 kilometers. At 8–12 minutes, communication speed is roughly 32 bytes per second. In about 15 minutes, analogue sensor signals are translated to digital and sent to the send-buffer. Each buffer packet takes 30–40 seconds for the transmitter to pass data from the buffer. On a daily basis, the system enables 12–14 satellites to communicate with each other at a rate of 32 bytes per contact. The satellite-based tracking device was designed to facilitate the examination of different parameters of renewable energy systems. Any signal from an analogue input must pass through an input multiplexer before being amplified by a differential amplifier. Noises are then removed using a low-pass filter. The machine has a maximum capacity of 12 volts. Figure 4.9 shows the average solar PV module waste factors such as damages, defects between cell connection, power cable junction, optical failure, etc. These failure rates can be gradually reduced by

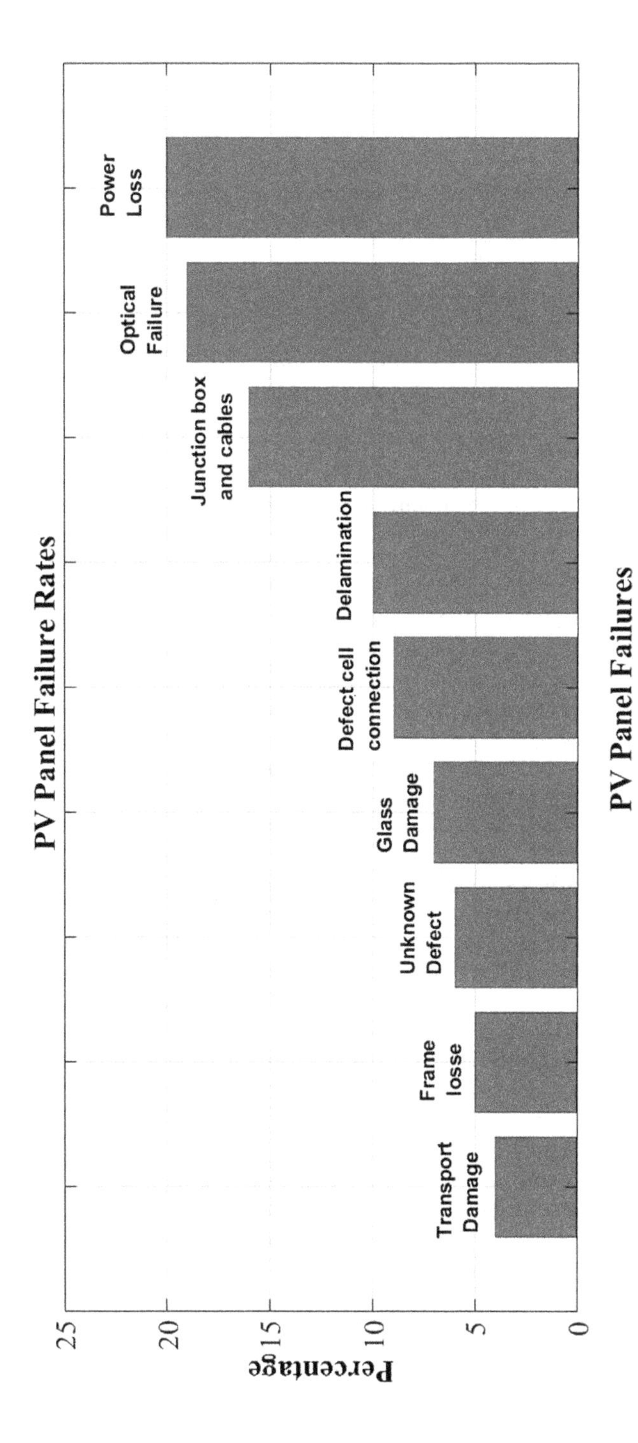

FIGURE 4.9 Estimated solar PV module waste.

implementing predictive emerging techniques in solar plant operation and maintenance. In the future, solar plants can be constructed with predictive maintenance system, which would greatly improve efficiency.

4.5 CONCLUSION

Predictive maintenance of a PV system has been discussed in this chapter in terms of installation, performance dependencies, various factors of maintenance in electrical system, inverter system, rooftop system, rack system, ground mounting system, tracking mount system, environment conditions, and its classifications.

REFERENCES

1. Gallardo-Saavedra, S., Hernández-Callejo, L., & Duque-Pérez, O. (2019). Quantitative failure rates and modes analysis in photovoltaic plants. *Energy*, 183, 825–836.
2. Adhya, S., Saha, D., Das, A., Jana, J., & Saha, H. (2016, January). An IoT based smart solar photovoltaic remote monitoring and control unit. In *2016 2nd International Conference on Control, Instrumentation, Energy & Communication (CIEC)* (pp. 432–436). IEEE.
3. Moreno-Garcia, I. M., Palacios-Garcia, E. J., Pallares-Lopez, V., Santiago, I., Gonzalez-Redondo, M. J., Varo-Martinez, M., & Real-Calvo, R. J. (2016). Real-time monitoring system for a utility-scale photovoltaic power plant. *Sensors*, 16(6), 770.
4. Hernandez, R. R., Easter, S. B., Murphy-Mariscal, M. L., Maestre, F. T., Tavassoli, M., Allen, E. B., & Allen, M. F. (2014). Environmental impacts of utility-scale solar energy. *Renewable and Sustainable Energy Reviews*, 29, 766–779.
5. Cameron, C. P., Boyson, W. E., & Riley, D. M. (2008, May). Comparison of PV system performance-model predictions with measured PV system performance. In *2008 33rd IEEE Photovoltaic Specialists Conference* (pp. 1–6). IEEE.
6. Cristaldi, L., Khalil, M., & Soulatiantork, P. (2017). A root cause analysis and a risk evaluation of PV balance of system failures. *Acta Imeko*, 6, 113–120.
7. Peters, L., & Madlener, R. (2017). Economic evaluation of maintenance strategies for ground-mounted solar photovoltaic plants. *Applied Energy*, 199, 264–280.
8. Ramu, S. K., Irudayaraj, G. C. R., & Elango, R. (2021). An IoT-based smart monitoring scheme for solar PV applications. In *Electrical and Electronic Devices, Circuits, and Materials: Technological Challenges and Solutions*, 211–233. https://doi.org/10.1002/9781119755104.ch12
9. Osmani, K., Haddad, A., Lemenand, T., Castanier, B., & Ramadan, M. (2020). A review on maintenance strategies for PV systems. *Science of the Total Environment*, 746, 141753.
10. Bosman, L. B., Leon-Salas, W. D., Hutzel, W., & Soto, E. A. (2020). PV system predictive maintenance: Challenges, current approaches, and opportunities. *Energies*, 13(6), 1398.
11. Suresh, K. P., & Ramesh, S. (2019). Grid-interconnected solar photovoltaic system for power quality improvement using extended reference signal generation strategy. *Journal of Testing and Evaluation*, 49, 1–21.
12. Pal, N. R., Pal, S., Das, J., & Majumdar, K. (2003). SOFM-MLP: A hybrid neural network for atmospheric temperature prediction. *IEEE Transactions on Geoscience and Remote Sensing*, 41(12), 2783–2791.
13. Krishnamurti, T. N., Kishtawal, C. M., LaRow, T. E., Bachiochi, D. R., Zhang, Z., Williford, C. E., ... & Surendran, S. (1999). Improved weather and seasonal climate forecasts from multimodel superensemble. *Science*, 285(5433), 1548–1550.

14. Lazos, D., Sproul, A. B., & Kay, M. (2015). Development of hybrid numerical and statistical short term horizon weather prediction models for building energy management optimisation. *Building and Environment*, 90, 82–95.
15. Möller, A., & Groß, J. (2016). Probabilistic temperature forecasting based on an ensemble autoregressive modification. *Quarterly Journal of the Royal Meteorological Society*, 142(696), 1385–1394.
16. Rahman, M. M., Selvaraj, J., Rahim, N. A., & Hasanuzzaman, M. (2018). Global modern monitoring systems for PV based power generation: A review. *Renewable and Sustainable Energy Reviews*, 82, 4142–4158.
17. López, M. E. A., Mantinan, F. J. G., & Molina, M. G. (2012, September). Implementation of wireless remote monitoring and control of solar photovoltaic (PV) system. In *2012 Sixth IEEE/PES Transmission and Distribution: Latin America Conference and Exposition (T&D-LA)* (pp. 1–6). IEEE.
18. Ben-Menahem, S., & Yang, S. C. (2012, June). Online photovoltaic array hot-spot Bayesian diagnostics from streaming string-level electric data. In *2012 38th IEEE Photovoltaic Specialists Conference* (pp. 002432–002437). IEEE.
19. Lin, C., Jie, G., Wu, Z., & Rui, W. (2011, August). Design of networked monitoring system of PV grid-connected power plant. In *Proceedings of 2011 International Conference on Electronic & Mechanical Engineering and Information Technology* (Vol. 3, pp. 1169–1172). IEEE.
20. Krauter, S., & Depping, T. (2003, May). Monitoring of remote PV-systems via satellite. In *Proceedings of 3rd World Conference on Photovoltaic Energy Conversion, 2003* (Vol. 3, pp. 2202–2205). IEEE.

5 Machine Learning–Based Predictive Maintenance for Solar Plants for Early Fault Detection and Diagnostics

S. Saravanan
Sri Krishna College of Technology, Coimbatore, India

P. Pandiyan
KPR Institute of Engineering and Technology, Coimbatore, India

T. Rajasekaran
KPR Institute of Engineering and Technology, Coimbatore, India

N. Prabaharan
School of Electrical & Electronics Engineering, SASTRA Deemed University, Thanjavur, India

T. Chinnadurai
Sri Krishna College of Technology, Coimbatore, India

Ramji Tiwari
Sri Krishna College of Engineering and Technology, Coimbatore, India

CONTENTS

DOI: 10.1201/9781003202288-5

LEARNING OUTCOMES

i. To understand the significance of fault detection and diagnosis in solar photovoltaic systems
ii. To apply machine learning techniques to detecting online faults in solar PV plants
iii. To understand the predictive maintenance analysis for real-time scenarios

5.1 INTRODUCTION

There has been a remarkable increase in the number of solar photovoltaic (SPV) plants/farms around the world in recent decades. This is mainly attributable to problems such as environmental pollution, global warming, and the depletion of nonrenewable energy sources. Nonrenewable resources like coal, oil, and natural gas have a negative impact on eco-biodiversity, climate change, and human health. By comparison, renewable energy resources do not produce carbon dioxide emissions and are eco-friendly, secure, and sustainable [1]. Masson et al. [2] reported that installing 100 GW of SPV plants in Europe by 2012 has prevented over 53 million tons of carbon dioxide emissions. SPV plants are built in order to withstand the effects of extreme weather, ultraviolet radiation, and wind shears on the panels. Still, damage and faults do occur and may result in lower power output, shortened lifetime of SPV panels, and safety hazards for plant operators [3].

In recent decades, researchers have been investigating the issues of fault identification and diagnosis of SPV plants [4]. These approaches can be categorized into traditional threshold assessment methods and machine learning (ML)-based methods [5]. In the traditional threshold method, identification and diagnosis of fault can be attained through electrical parameters such as voltage (v), current (I), and power output. The research work on traditional threshold approaches is discussed in [1]. Chouder et al. [6] proposed an effective method of monitoring and fault identification of SPV plants using power loss analysis with four electrical indicators such as thermal capture losses, voltage ratio, current ratio, and miscellaneous capture losses. Drews et al. [7] suggested a remote monitoring and fault detection using solar radiance data received from a satellite rather than using actual real-time data. A fault is declared when there is a significant difference between the measured output value and the simulated power data. Authors [1] suggested a fault identification (detection) and diagnosis method using an exponentially weighed moving average monitoring map to detect three kinds of faults, namely temporary shading, short-circuit (SC), and open-circuit (OC) faults. This approach calculates maximum power point residuals of voltage, current, and output power based on real-time data and the numerical simulation data (fault-free data). Traditional threshold evaluation methods are simple to perform and can produce accurate results; their main drawbacks is that calculations must be done manually and that the SPV arrays must operate at an optimum power point.

ML-based techniques are the second form of fault identification and diagnosis in SPV plants/systems. In order to identify and diagnose the eventual faults, ML-based approaches need to gather the healthy and faulty data from an operating SPV system. The literature based on the machine learning approach is discussed in [8]. Karatepe et al. [9] detected and located the SC faults in the panels of an SPV array using a three-layered artificial neural network (ANN) architecture. Zhao et al. [10] suggested several methods for detecting and diagnosing various faults in a SPV system, including OC faults, SC faults, partial shading occurrence, degradation fault, and line-to-line (L-L) fault, using decision tree and graph-based semi-supervised learning approaches and outlier detection laws. The measured voltage (v) as well as current (I) values of each individual SPV panel are required by the outlier detection law approach, which makes it a very expensive technique. Furthermore, in order to generate an effective training process, the decision tree method requires a large amount of high-quality data.

In [11], an ANN approach was used to identify and classify eight different types of SPV module failures. In these studies, ANN was used in conjunction with other conventional threshold assessment methods to identify and diagnose the faults in SPV systems. A traditional threshold method was used for fault detection, while an ANN classifier was used to classify the faults. The ANN approach of the multilayer perceptron (MLP) kind is used in both studies, and they are trained iteratively using the gradient descendant algorithm (DGA) to estimate the back-propagation difference between actual outputs and goals. Due to its slow training steps, this form of network is more likely to fall into local minima rather than global minima. These issues affect the system's dependability and performance [5]. Furthermore, this type of approach needs a dataset that contains a huge amount of high-quality labeled data that accurately defines the process [5].

In [12], probabilistic neural networks have been identified as an effective method for solving classification problems. This approach has several significant advantages, including fast computation and ease of use during the training phase because of lack of weight-adjustment requirements. Furthermore, this approach can identify every new incoming signal without carrying out the entire training phase. Specifically, the probabilistic neural network classifier has a high level of robustness against noisy input data provided by sensors and other measuring instrument data.

A fault identification and diagnosis method for a grid-connected SPV plant is proposed in [13], which is based on the Probabilistic Neural Network (PNN) classifier technique. In this research work, a valid model for a real-time operating SPV system, is obtained employing a single diode model as well as a parameters extraction method. Once the proposed model gets validated by testing, faulty data samples are collected, which are utilized as input data for the PNN classifier. This method employs two PNN network classifiers. The first classifier is used to locate the existence of faults, whereas the second is used to detect the type of fault.

In general, the fault identification and diagnosis approach is divided into four stages, namely extraction of SPV panel parameters using an algorithm, model validation, related dataset collection, and identification and diagnosis of faults using classifiers. This chapter focuses on prediction system failures by identifying the anomalies in SPV plant/system prior to causing substantial damage.

5.2 MACHINE LEARNING TECHNIQUES

It is essential to add some principles about the development of a machine learning (ML) model before presenting the review's findings on a predictive maintenance (PrM) application. There are several steps involved, according to Soares [14], including relevant data collection, data preprocessing, model selection phase, model training phase, model validation, and model maintenance. These measures are depicted in Figure 5.1.

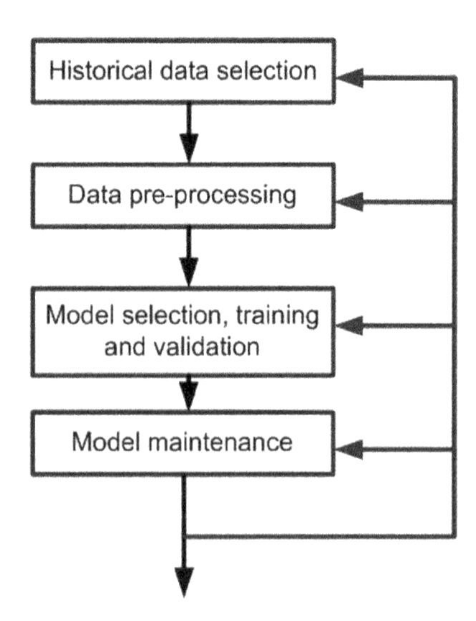

FIGURE 5.1 Procedure to design the machine learning model.

The relevant data collection phase determines how data are gathered and processed in order to select useful data for the ML algorithm. Jardine et al. [15] reported that this phase is also known as data acquisition in the PrM application, and it seeks to gather necessary data about the system's health. In the data preprocessing phase, data are processed and transformed so that the ML model can process it efficiently. This phase consists of data transformation, data cleaning, and data reduction. The data preprocessing phase in PrM addresses and investigates the gathered data for a better interpretation.

The model selection phase is used to choose an appropriate ML model. In the model training stage, model development of the ML algorithm takes place. The validation phase is used to determine the feasibility of a proposed model to represent the underlying system. This phase is also known as the maintenance decision-making phase in PrM, and it is responsible for choosing the best algorithm for the PrM application. The objective of the model maintenance phase is to keep the model performance over time. Industrial applications are likely to change over time, causing the model to perform differently. As [14] recommended future research look for more details on techniques that can be used in each phase.

The researchers use machine learning algorithms such as artificial neural networks (ANN), fuzzy logic (FL), and others to develop the possible methods of detecting the faults in SPV systems. Machine learning algorithms are trained and tested using model or experimental SPV data values to learn the association with output as well as input parameters of a SPV system to detect the fault. Anomaly data owing to fault events are also needed along with normal SPV data for accurate training. Once the instantaneous SPV data (irradiation levels and temperature) are sensed and provided as inputs to the algorithm, the trained ML model will accurately detect both ideal and faulty conditions in real time. As a result, using simple error assessment, any kind of fault conditions of the SPV system can be easily detected. Figure 5.2 depicts the flowchart for ML-based SPV fault detection [16].

5.3 PV PLANT MODEL

A SPV system is expressed by Equation 5.1

$$P = f\left(I_r, T\right) \tag{5.1}$$

where P represents power output (kW), T stands for environmental temperature (°C), and I_r denotes solar irradiance (W/m^2).

In general, an artificial neural network (ANN) is used to derive a useful function from Equation 5.1, which has the property that it learns from an available data set and the system is nonlinear. Therefore, the ANN technique can be adopted using the Multi-layer Perceptron concept [11] with one output (P) and two inputs (I_r and T).

The datasets are collected from actual power generation plants in which unwanted data are filtered and removed in order to feed the data to train the ANN network. The data set should be labeled correctly, otherwise it affects the predictive model.

A dataset should be collected from an actual power plant for training the network. However, the data set must be carefully filtered to exclude incorrect readings that

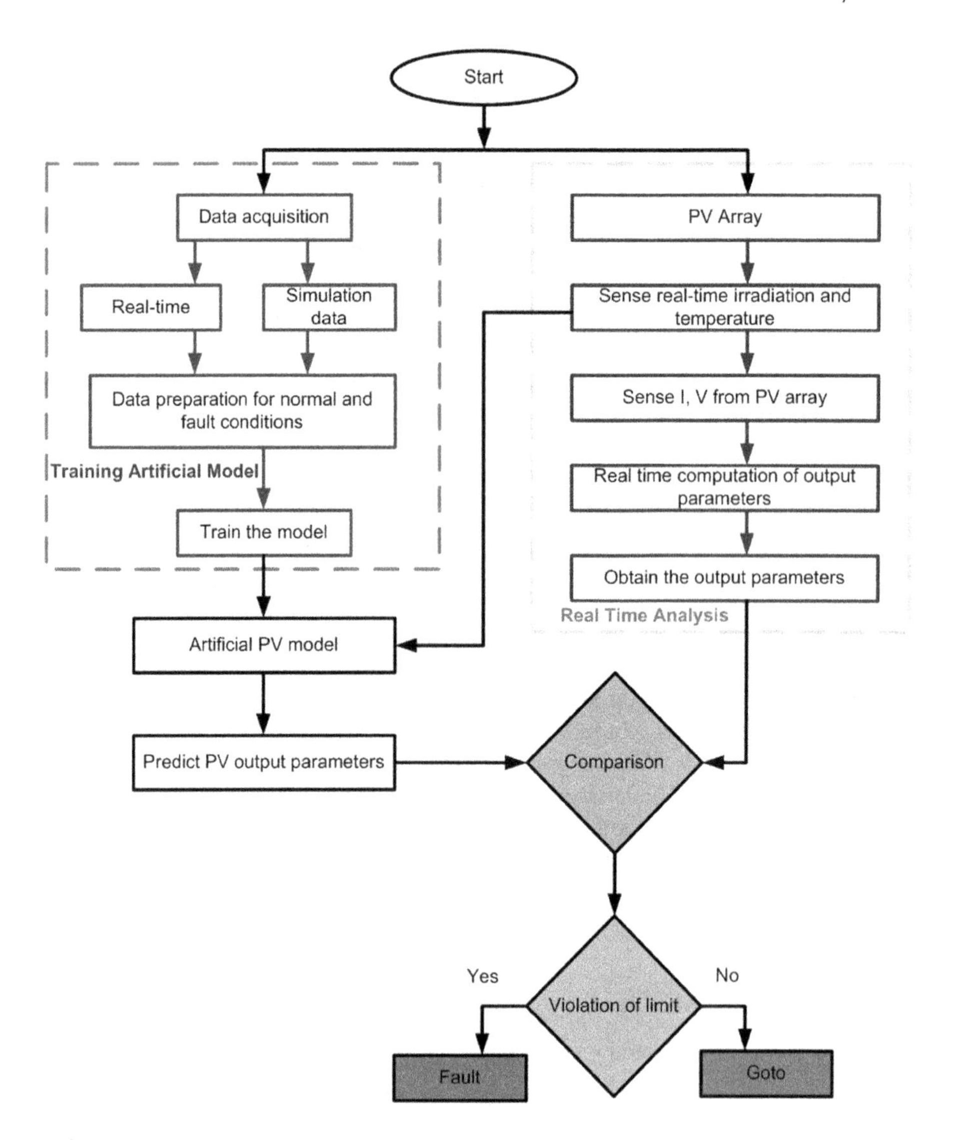

FIGURE 5.2 Flowchart for fault detection using machine learning methods.

could have an effect on the predictive model. Constructing the best model needs a consistent training data set. The available data set value for the training must contain and reflect all seasonal conditions (at least one year's worth of data). To ensure that all collected data are accurate, a preprocessing procedure must be performed on the available data. In practice, available data that show circumstances where output power is close to zero but the radiance value is somewhat greater than zero is denoted by a red oval. This event is referred to as plant failure or plant maintenance in which the DC-to-AC converter (i.e., inverter) is in an "OFF" condition. These kinds of data should be excluded because they do not reflect standard operating conditions. Data indicating patterns associated with extremely low levels of sunlight (zero output power and low radiance profile) should also be omitted during this filtering operation, as this reduces the measurement accuracy significantly.

In the second preprocessing stage, averaging the data over one hour is done. Data collected from SPV systems is generally sampled over a shorter duration (e.g., 1–10 min), although data with fixed sampling time may be highly inconsistent. The averaging of the collected data over one hour can thus reduce dataset variability and contribute to improving the built model accuracy. Computing a mean over a defined time interval is an activity that allows maintaining the regular solar process core variance when dealing with slow-changing signals like solar radiance. In this process, usage of time periods other than one hour may lead to missing the appropriate data. Alternatively, using a small time window for mean estimation will make it more difficult to extrapolate long-term trends.

5.4 PV PREDICTIVE MAINTENANCE

5.4.1 Manual Diagnostics–Based Predictive Maintenance

This approach is the cheapest in general, but its prediction accuracy is the lowest. In this section, both qualitative approaches such as visual examination of individual parts and whole SPV plant and infrared (IR) thermography of entire SPV modules and quantitative methods like V-I characteristics and insulation resistance analysis of SPV module are discussed for performing predictive maintenance.

To begin, a visual inspection of the individual parts and the whole system may be performed to see if any possible problems exist. As discussed in [17], shade analysis is performed using a Solmetric Suneye tool to detect possible mismatch problems caused by new construction or/and tree growth. According to [18], visual manual inspection may also be carried out to detect soiling on the SPV modules, such as bird droppings, fallen leaves, dust, snow, and other residues. However, visual inspection presents problems and opportunities, one of which is the safety of identification and cleaning.

In the next stage, IR thermography is used to determine the efficiency of SPV modules on a qualitative level. For large SPV plants, this method is always combined with intelligent drones or unmanned aerial vehicles (UAVs) as reported in [19]. As presented in [20], aerial IR thermography is a well-known diagnostic approach for assessing the output of SPV modules with less time and greater accuracy, but it is a more expensive technique.

It is worth noting, however, that a combination of visual cameras and thermal imaging has been suggested as a cost-effective and reliable approach [21]. In most SPV module production processes, thermal imaging is already in use to identify a number of probable issues. The use of the similar qualitative method in a real-time environment has therefore been much praised, even though pass/fail standards are yet to be discussed [22]. As mentioned by [23], thermal imaging has been used to detect cell fractures, soldering defects, substrings that have been bypassed, and cells that have short-circuited. Researchers have suggested an automated method for SPV module monitoring in some situations, which utilizes image-based automatic detection of faults and defects as well as steps to correct the situation [24]. It is also worth noting that aerial thermal imaging faces some challenges and opportunities, such as quality of image, which is heavily affected by the unmanned aerial vehicle's (UAV's) observation angle, altitude, and velocity, as well as a more comprehensive understanding of qualitative and visual requirements.

In the last step, V-I characteristics and insulation resistance analysis of SPV modules will identify possible performance problems within individual modules from a quantitative viewpoint. Well-trained and experienced installers prefer a Solmetric PVA-1000S PV Analyzer Kit to measure the V-I characteristics of the SPV modules, and several other instruments such as voltage meter, amp meter, and power meter to validate the desired output of the other elements. The Solmetric analyzer is a portable device that is specifically developed to evaluate the performance of SPV panels during installation and commissioning and recommissioning as well as troubleshooting. The conduct of PV panels can be examined using an I-V curve tracer, which takes into account ambient temperature and solar irradiance. Fault detection and short circuit bypass diodes can also cause irregularities that can be easily identified with an I-V curve tracer. According to [25], the time required to determine the health of SPV panels using an I-V curve tracer is calculated to be as little as one minute per panel. While there are numerous advantages to manually assessing the output of SPV modules with handheld instruments, challenges and opportunities remain due to the time taken to evaluate an entire plant, as well as the possibility of user mistakes in the data collection process.

In conclusion, when all these approaches are combined, it is more likely that problems will be detected. However, there are always problems and opportunities to be had. These methods are mainly utilized to evaluate the efficiency of SPV modules and do not take the SPV system as a whole into consideration.

5.4.2 FMEA-Based Predictive Maintenance

The failure mode and effects analysis (FMEA) method in general is moderately costly and provides a medium level of identification accuracy. The FMEA is a semiquantitative approach to preventing failures and analyzing the risks by determining causes and consequences of the system in order to assess the steps that can be taken to prevent failures. This method was applied to support fault in design phase, testability, logistic support, protection, and other related functions in the nuclear, semiconductor, automotive, and aerospace industries. In SPV systems, the FMEA method has been utilized to classify the parts/elements with the greatest chances of failure. While some studies focused on an analysis of historical data to find out the components/parts with the highest chances of failure, other reviewed works centered on SPV plants exposed to particular environmental conditions. The studies have revealed a variety of outcomes, but all have the similar goal of diminishing the risk of SPV systems failure and improving their efficiency.

As presented in [26], the first step in developing a SPV system's predictive life model is to detect the failures of its components. The failure of components associated with the SPV system has two key consequences: (1) the parts are damaged, and (2) energy generation diminishes. Researchers [27] developed the extension of FMEA analysis (FMECA-failure mode effects criticality analysis) with the Markov process to define and prioritize the critical SPV system's components. In the process, they have discovered that the inverter in the SPV system has a higher risk priority number compared to other components, requiring that inverter maintenance be prioritized. Colli [28], who used the FMEA approach to analyze the SPV system configuration at the Brookhaven National Laboratory, came to the same conclusion that the inverter as well as the ground system have the highest likelihood of failure in a SPV plant/farm. Villarini and Cesarotti [29] improved the FMEA method with actual SPV farm data, climatic conditions, and maintenance technicians' opinions on the predictive maintenance analysis and found that the inverter and the ground system have the highest risk priority number; they also reported that component failures were caused by a system overload and a lack of isolation.

SPV panel installations in desert climates have also been subjected to the FMEA approach. Studies [26] reported on SPV data collected from solar plants in Arizona and found that encapsulation discoloration and solder bond fatigue were the leading causes of failure. Shrestha et al. [30] suggested that the deterioration of the solder bond could result in hotspots or back sheet burns, which could compromise the PV system's protection.

The FMEA approach aims to reduce the effect of potential SPV system failures and thus enhance the electrical efficiency. On the other hand, FMEAC technique has shown that the presence of dust and contamination on the SPV module surfaces reduces the SPV system's efficiency [31].

Catelani and Ciani [31] used a link between energy reduction and MPP of the SPV plant to validate the output of the FMECA approach using a sensitivity study of the MPP and dust concentration to optimize the electrical efficiency of the SPV plant. Authors [32] analyzed the failure rates and 63 PV plant modes quantitatively. The study revealed that the monitoring systems, inverters, grid, and communication systems are the elements with the greatest risk of failure in SPV plants. Furthermore, the failure rate/unit of power was found to be higher in small plants. The annual failure rate for SPV systems with less than 750 kW installed is more than double that that of larger plants. However, even with the SPV plants that have been in operation longer, the research work concluded that careful maintenance will keep the defect rate constant. The probabilities of failure of SPV plant components are an important input to the predictive maintenance models. As a result, considering the possibility of component failure is an important aspect of improving predictive maintenance models. It would be possible to extend the life span of SPV systems and reduce the likelihood of failure in this way.

5.4.3 Machine Learning and Forecasting Method–Based Predictive Maintenance

This method is fairly costly and provides a medium level of prediction accuracy. Recent solar energy research has looked at localized forecasts for short-time periods using microclimate parameters and regressive processes, finding the parametric factors with higher contribution rates, and uncovering temporal trends in weather data. In some research works, autoregressive integrated moving average (ARIMA) models, ANN, and other methods have been used to enhance the prediction accuracy by considering factors such as wind velocity, time of day, atmospheric pressure, and cloud cover. Studies of additional parameters have been applied to determine SPV output using ANN technique.

Different types of estimation models such as stochastic learning methods, numerical weather prediction models, ARIMA modeling, and time series neural networks have been used to predict solar irradiance data. Researchers [33] found that inputs identified by a combination of genetic algorithm (GA) and Gamma test were optimal and provided the best output when fed to a neural network based on research work [34]. In comparison to models of reference such as Perez's and Persistent models, this combination resulted in a 10–15% increase in Root Mean Square Error (RMSE). In addition, increasing the forecasting horizon resulted in less accuracy. But due to lengthy daytime during the summer season, this lost accuracy was less pronounced than during the winter season, which is typically cloudier. Another method for predicting solar irradiance was to use a mesoscale weather prediction model numerically as reported in [35], which measured the solar radiance and found that the model's accuracy was dependent on the air's aerosol optical depth. The accuracy of solar radiance forecast models increases when measurements of aerosols and their assimilation depth are considered. Improved understanding of aerosol physics in estimation models is thus another field to focus on when forecasting weather parameters used to measure solar energy.

5.4.4 Real-Time Sensors–Based Predictive Maintenance

This is the most expensive approach in general, but it has the greatest prediction accuracy. According to the discussion in [36], sensors were utilized for monitoring the SPV system in which the SPV module automatically orients itself with respect to the direction of the sun. Researchers have been using sensors to track the SPV system behavior in real time in some situations to evaluate the quality control issues. A low-cost SPV system monitoring proposed in [37] utilizes the sensors to read SPV module and ambient temperature, incoming solar radiance, SC current, OC voltage, fill factor, parameters connected with maximum power, and SPV module performance. In this study, researchers used a thermocouple to detect temperature, a pyranometer to measure solar radiance, a voltage divider to obtain voltage, and a shunt resistor to measure current. In addition, software applications were developed using LabView [38] and Visual Basic. The Arduino was used as a microcontroller/analyzer with software applications, which was compared with the commercially available Solmetric I–V curve tracer. Algorithms for detecting faults may be used to equate the predicted performance to real output using a series of errors thresholds based on fault-free systems reported in [39], allowing for the detection of both total and partial failure productivity.

Prieto et al. [40] presented the work on real-time monitoring of SPV system with wireless sensor networks and also analyzed the tracking performance, weaknesses, features, and failures. The collected data were analyzed with the aid of MATLAB® software to increase the productivity and energy output of SPV plants. Adhya et al. [41] used the Internet of Things (IoT) framework to monitor, collect real-time data, detect the fault, and conduct preventive maintenance of an SPV system by incorporating wireless sensors and transferring real-time data to a web application. Shariff et al. [42] designed the hardware (microcontroller) as well as the software (graphical user interface) for monitoring SPV systems through a GSM modem for real-time data transmission. SPV systems can be remotely controlled using GSM, in turn increasing the reliability and removing the risks of data transmission through wireless technology.

5.5 FAULTS DETECTION AND DIAGNOSIS STRATEGY

This section discusses design and development of an effective and reliable method for detecting and diagnosing faults in SPV systems using a PNN classifier. This method requires a high-quality database for identification and classification problems. In practice, collecting such a database is not always possible, particularly in SPV systems. Indeed, operating a SPV module in the presence of certain kinds of failures will render the system fully insecure, causing safety risks and potential catastrophic damage. As a result, the most effective way to address this issue is to provide a trusted model that accurately simulates the actual behavior of a SPV system in both faulty and healthy states. The flow diagram in Figure 5.3 illustrates the steps to be followed to detect and diagnose the faults in the SPV system. These include parameter extraction of SPV panels, model validation through experiment, database creation, fault identification, and diagnosis.

5.5.1 PV Module Parameters Extraction

In general, the single diode model is preferable for characterizing SPV panels. The output voltage (V_{SPV}) versus output current (I_{SPV}) in this model is expressed in Equation 5.2 [1]:

$$I_{SPV} = I_{\lg} - \underbrace{I_{sat}\left(\exp\left(\frac{q(V_{SPV}+R_{se}I_{SPV}}{nk_BT}\right)-1\right)}_{I_d} - \underbrace{\frac{V_{SPV}+R_{se}I_{SPV}}{R_{sh}}}_{I_{sh}} \tag{5.2}$$

where V_{SPV} represents the SPV panel's voltage, I_{SPV} denotes the SPV panel's current, I_{lg} stands for Sun light generated current, I_{sat} corresponds to diode saturation current, Rsh and Rse are the shunt and series resistance correspondingly, n represents the diode ideality factors, T corresponds to the SPV panel temperature, k_B denotes the Boltzmann constant, and q represents the charge.

The single diode model parameters extraction phase is an optimization problem in which cost criteria should be optimized using measured and estimated current as in Equation 5.2.

$$RMSE = \sqrt{\frac{1}{S_Z}\sum_{i=1}^{S_Z}\left[g_i\left(I_{md},V_{md},\gamma\right)\right]^2} \tag{5.3}$$

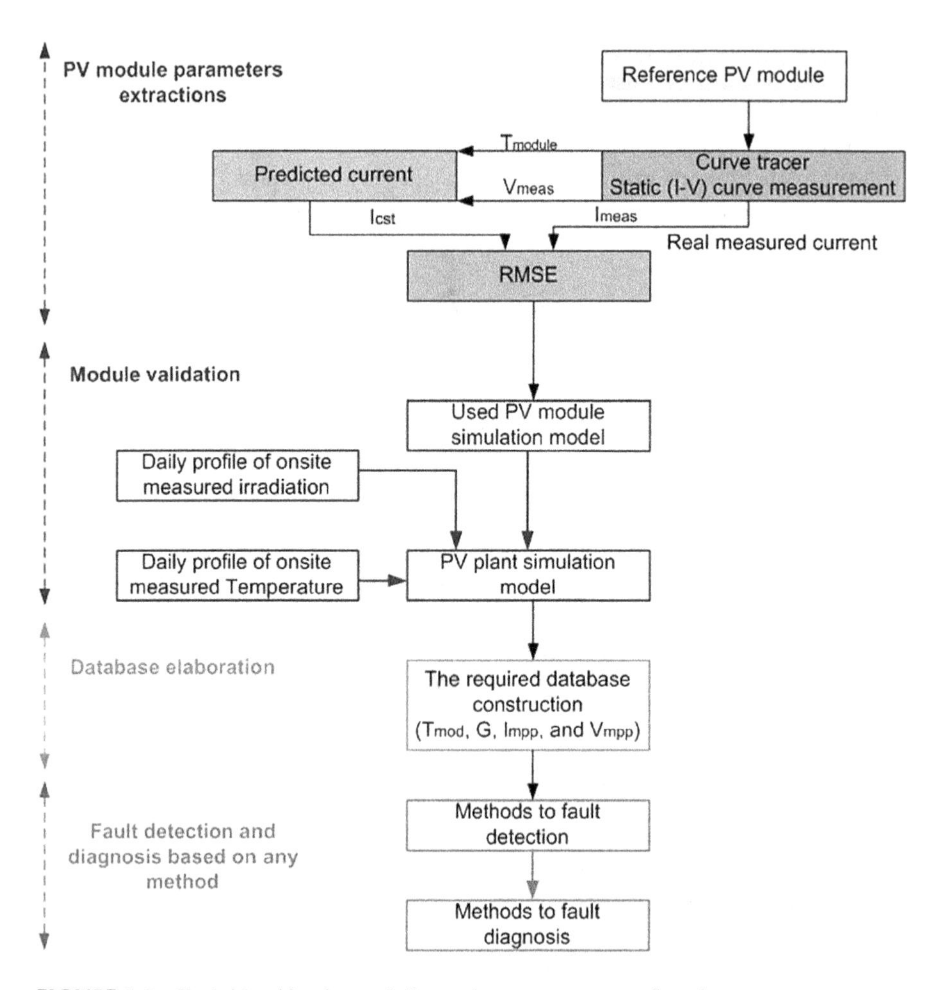

FIGURE 5.3 Fault identification and diagnosis strategy process flowchart.

where, V_{md} and I_{md} represent the measured voltage and current of SPV panel, respectively, γ denotes SDM vector parameter, and S_z corresponds to the size of I-V experimental characteristics data utilized at the time of SDM parameters extraction phase. Figure 5.4 depicts the process of extracting SDM parameters. PV module temperature is denoted by the temperature used in this process. The primary aim of this phase is to identify the best SDM parameters (I_{lg}, I_{sat}, n, R_{sh}, R_{se}) for minimizing costs and producing the lowest RMSE value. The best algorithm so far to extract ODM parameters is the ABC algorithm that is a more accurate optimization problem solver.

5.5.2 Validation of the Model

The extracted SDM parameters are then utilized for simulating the actual SPV systems using MATLAB/PSIM co-simulation under normal conditions. In practice, the physical model of the solar panel is incorporated into PSIM in order to simulate the physical

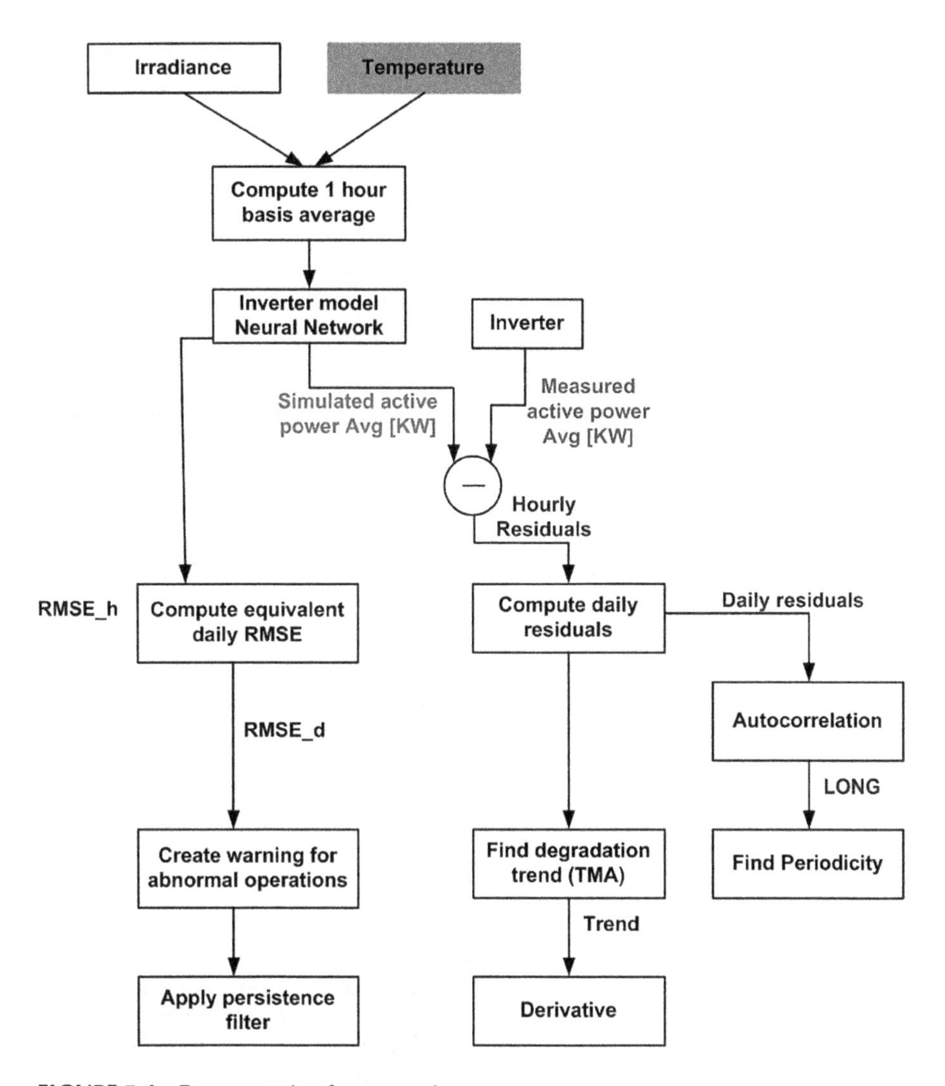

FIGURE 5.4 Representation for approach.

behavior of an actual SPV system. This can be accomplished by incorporating identified parameters of SDM into the physical model and then simulating the entire SPV system for real-time radiance and temperature measurements on a daily basis. Data processing as well as calculations are carried out on the Matlab/Simulink platform. In the last step, the measured power (P_{md}) and simulated power (P_{sd}) are compared.

5.5.3 Expansion of the Database

This phase consists of developing a high-quality repository (database) that provides a complete and realistic description of the SPV system under both normal and abnormal (faulty) circumstances. For this purpose, the validated SPV system model is utilized to create faulty and healthy samples by deliberately adding the desired faults in the real-time data, namely radiance profiles (G), panel temperature (T), voltage (V_{mp}), and current (I_{mp}) at maximum power point conditions.

5.5.4 Using a PNN Classifier to Identify and Diagnose Faults

The final step involves realizing two PNN classifiers based on the collected database. The first category is responsible for fault identification, whereas the second category is dedicated to detecting the fault root. The following are detailed explanations of the PNN architecture, training, and testing methodologies.

5.6 ANOMALY IDENTIFICATION AND PREDICTIVE WARNINGS USING RESIDUAL CONTROL

In this section, the algorithm related to anomaly identification and generating the predictive maintenance alert are discussed. As shown in Figure 5.4, this method is based on analyzing the residual vector, calculated by finding the difference between the ANN model output and the real-time SPV plant data. As the algorithm was developed to identify the data pattern, a sufficient amount of data had to be collected. With this objective, continuous data are supplied. Hence, older data are periodically removed when new data are collected and sampled from the plant, then the collected data series are utilized for computing the residual vector, which is passed to another analysis block that is explained as follows.

5.6.1 Making Daily Residuals Calculations

A per-day aggregation is the first procedure performed on the residuals vector. SPV data are typically sampled on a minute-by-minute basis, but a per-hour aggregation can also be considered as in [43]. Although a one-hour-based data collection is needed to develop the best plant model, the resultant data set may have too many data values for pattern-based anomaly identification. Analysis is based on identifying the data patterns that emerge with a less frequency in the order of days, but a one-hour-based data collection holds high frequencies that are not only unusable for further analysis but may also be harmful to the plant. As a result, beginning with the one-hour residual vector to calculate the daily vector by taking into account the total value of the absolute hourly residual vector data per day. Thus, the new data sequence includes a value for each day equal to the total of the absolute values of the remaining parts of the day.

Once the daily residuals are estimated, two distinct data processing ways are there as in Figure 5.4. In the first path, detecting "out of normal operation" conditions through threshold value as seen in the initial calculation generates alerts, while in another path, so called long-term analysis follows to discover potential wear and and tear (e.g., finding regions with increased abnormal residuals). The resulting predictive alerts are then combined and filtered. The following subsections go into the specifics of this procedure.

5.6.2 Determining Standard Operating Limits

A typical operating condition means that the plant's behavior does not deviate significantly from the expected one, resulting in low values in the residuals vector. The threshold value should be fixed based on daily residual value and generate an alert when a value exceeds those thresholds.

To validate the model predictions, RMSE value is used as base value. This RMSE (σ_h) value denotes the standard deviation of the usual activity of the AC power model and the model to be constructed using the hourly based data set. The two threshold limits are defined as hourly lower limit (H_{LL}) and hourly upper limit (H_{UL}). The hourly lower limit (H_{ll}) can be expressed in Equation 5.4

$$H_{LL} = 3\sigma_h,\ H_{UL} = 5\sigma_h \tag{5.4}$$

where H_{UL} represents the hourly upper limit.

Data samples that exceed H_{LL} value but fall short of H_{UL} indicate irregular SPV system activity (abnormal behavior), while the data samples beyond the value of H_{UL} indicate strong anomalies with a high degree of confidence. According to reference [44], when collected data is presented with a Gaussian distribution, around 99.7% of the data are within $3\sigma_h$, and around 99.99% of the data are within $5\sigma_h$. In addition, Chebyshev's theorem states that at least 88.9% of the observations would fall within $3\sigma_h$. As a result, $3\sigma_h$ as well as $5\sigma_h$ denote reasonable values for two threshold limits.

As mentioned earlier, anomaly detection is done through daily cumulative residual vectors. According to this assertion, two equivalent threshold values (H_{LL} and H_{UL}) are applied to daily data. The daily equivalent standard deviation index can be expressed using daily upper limit (D_{UL}) and lower limit (D_{LL}) as in Equation 5.5

$$\begin{aligned} \sigma_d &= \mathrm{K}\sigma_h \\ D_{LL} &= 3\sigma_d = 3(\mathrm{K}\sigma_h) \\ D_{UL} &= 5\sigma_d = 5(\mathrm{K}\sigma_h) \end{aligned} \tag{5.5}$$

Obviously, there is a direct correlation between the value of K and the accuracy of the system's predictive alerts. In practice, the lower the value of K, the lower the daily normal activity limits, and the greater the chance of obtaining daily residuals beyond those limits, resulting in a greater number of alerts and, as a result, a greater chance that any of them would be false positives. Inversely, the higher the value of K, the higher the daily normal activity limits, and the less likely daily residuals will reach those limits, increasing the risk of losing information possibly relevant to a specific anomaly. Furthermore, the amount of power generated by any SPV system can vary during the day due to a variety of factors such as variations in the solar radiation intensity throughout the day and the angle at which the sun's rays strike the earth's surface. Likewise, due to the impact of the different air mass density, insolation would be poor at high latitudes.

5.6.3 Identifying Residual Patterns

To improve the prediction accuracy in recognizing abnormal activities, the alerts provided in the earlier phase should be processed further, reducing the possibility of producing false positives and aiding in the computation of predictive maintenance warnings. In this case, the control quality parameters are focused on extracting the patterns from daily residual vectors and integrating them with alerts.

Study results obtained in [44] can be used to derive the trend in daily residual values depending on the filtered data using the triangular moving average (TMA) concept. The size of the window must be considered in the average calculation, which is the key parameter to be fixed for TMA. This can be calculated by looking for periodicity in residuals. In order to achieve this, the autocorrelation feature is used in a traditional data processing algorithm.

5.6.4 Creating Predictive Warnings

The final step in the data analysis phase is the generation of predictive warnings, which is accomplished by calculating the trend signal's derivative to identify areas of deterioration or convergence relative to the inverter's normal characteristics. Positive derivatives sample values represent degradation areas. On the other hand, negative values represent convergence or normal activity. Derivative samples are compared to residual warnings. Predictive maintenance warnings are produced only for residual alerts that correspond to positive trend derivative values. The trend is the result of a processing step that employs a time window and TMA. The size of the window included in this stage is related to the warning expectation. The data processing chain used in the process can provide predictive warnings with a high expectation, thus enabling maintainers to determine whether to schedule proper actions, since the windows size is in the order of one month in general.

5.6.5 Alert Usage and Decision-Making

The aim of the data analysis method is to detect a deterioration in performance that may lead to immediate failure. In this case, the conditional statement is necessary, because it is important to remember that the existence of an alert does not inherently mean the presence of a fault. In this phase, an "emergency indicator" is provided in the control room, which indicates that the SPV system's behavior varies slightly from the usual, but it is for operators to determine whether or not a maintenance procedure on a plant is to be planned. For example, an operator may decide to perform a maintenance action based on their experience when a few numbers of repeated alerts are informed. This value can be determined using a variety of factors, including previous alerts and their correlation to occurred faults, the age of the inverters or plant's SPV modules, and so on, but it is ultimately determined by the operator's decision.

5.7 CONCLUSION

In this chapter, machine learning algorithms for fault identification and diagnosis of SPV systems are presented. Fault detection and diagnosis consist of four main phases, namely parameter extraction of SPV module using different algorithms (ABC algorithm is the best so far), simulation of the entire system using simulation software (PSIM-MATLAB), high-quality database creation with faulty and healthy operations, and effective fault identification and diagnosis based on a classifier algorithm (PNN classifier is the best so far). In order to provide daily predictive warnings to SPV system/plant operators for supporting the maintenance decision process, anomaly identification and predictive maintenance model are also discussed.

REFERENCES

1. Garoudja, E., Harrou, F., Sun, Y., Kara, K., Chouder, A., & Silvestre, S. (2017). Statistical fault detection in photovoltaic systems. *Solar Energy*, 150, 485–499.
2. Masson, G, Latour, M, Rekinger, M, Theologitis, I-T, & Papoutsi, M. (2013). Global market outlook for photovoltaics 2013–2017. *European Photovoltaic Industry Association*, 12–32.
3. Daliento, S., Chouder, A., Guerriero, P., Pavan, A. M., Mellit, A., Moeini, R., & Tricoli, P. (2017). Monitoring, diagnosis, and power forecasting for photovoltaic fields: A review. *International Journal of Photoenergy*, 2017, 1–13.
4. Spagnuolo, G., Xiao, W., & Cecati, C. (2015). Monitoring, diagnosis, prognosis, and techniques for increasing the lifetime/reliability of photovoltaic systems. *IEEE Transactions on Industrial Electronics*, 62(11), 7226–7227.
5. Chen, Z., Wu, L., Cheng, S., Lin, P., Wu, Y., & Lin, W. (2017). Intelligent fault diagnosis of photovoltaic arrays based on optimized kernel extreme learning machine and IV characteristics. *Applied Energy*, 204, 912–931.
6. Chouder, A., & Silvestre, S. (2010). Automatic supervision and fault detection of PV systems based on power losses analysis. *Energy conversion and Management*, 51(10), 1929–1937.
7. Drews, A., De Keizer, A. C., Beyer, H. G., Lorenz, E., Betcke, J., Van Sark, W. G. J. H. M., ... & Heinemann, D. (2007). Monitoring and remote failure detection of grid-connected PV systems based on satellite observations. *Solar Energy*, 81(4), 548–564.
8. Bosman, L. B., Leon-Salas, W. D., Hutzel, W., & Soto, E. A. (2020). PV System predictive maintenance: Challenges, current approaches, and opportunities. *Energies*, 13(6), 1398.
9. Karatepe, E., & Hiyama, T. (2011, September). Controlling of artificial neural network for fault diagnosis of photovoltaic array. In *2011 16th International Conference on Intelligent System Applications to Power Systems* (pp. 1–6). IEEE.
10. Zhao, Y., Ball, R., Mosesian, J., de Palma, J. F., & Lehman, B. (2014). Graph-based semi-supervised learning for fault detection and classification in solar photovoltaic arrays. *IEEE Transactions on Power Electronics*, 30(5), 2848–2858.
11. Chine, W., Mellit, A., Lughi, V., Malek, A., Sulligoi, G., & Pavan, A. M. (2016). A novel fault diagnosis technique for photovoltaic systems based on artificial neural networks. *Renewable Energy*, 90, 501–512.
12. Specht, D. F. (1990). Probabilistic neural networks and the polynomial adaline as complementary techniques for classification. *IEEE Transactions on Neural Networks*, 1(1), 111–121.
13. Garoudja, E., Chouder, A., Kara, K., & Silvestre, S. (2017). An enhanced machine learning based approach for failures detection and diagnosis of PV systems. *Energy Conversion and Management*, 151, 496–513.

14. Soares, S. G. (2015). Ensemble learning methodologies for soft sensor development in industrial processes. Ph.d. thesis Computer engineering, faculty of sciences and technology. Coimbra, Portugal: Department of Electrical and University of Coimbra.
15. Jardine, A. K., Lin, D., & Banjevic, D. (2006). A review on machinery diagnostics and prognostics implementing condition-based maintenance. *Mechanical Systems and Signal Processing*, 20(7), 1483–1510.
16. Pillai, D. S., Blaabjerg, F., & Rajasekar, N. (2019). A comparative evaluation of advanced fault detection approaches for PV systems. *IEEE Journal of Photovoltaics*, 9(2), 513–527.
17. Bulanyi, P., & Zhang, R. (2014, May). Shading analysis & improvement for distributed residential grid-connected photovoltaics systems. In *The 52nd Annual Conference of the Australian Solar Council.*
18. Figgis, B., Ennaoui, A., Ahzi, S., & Rémond, Y. (2016, November). Review of PV soiling measurement methods. In *2016 International Renewable and Sustainable Energy Conference (IRSEC)* (pp. 176–180). IEEE.
19. Grimaccia, F., Aghaei, M., Mussetta, M., Leva, S., & Quater, P. B. (2015). Planning for PV plant performance monitoring by means of unmanned aerial systems (UAS). *International Journal of Energy and Environmental Engineering*, 6(1), 47–54.
20. Gallardo-Saavedra, S., Hernández-Callejo, L., & Duque-Perez, O. (2018). Technological review of the instrumentation used in aerial thermographic inspection of photovoltaic plants. *Renewable and Sustainable Energy Reviews*, 93, 566–579.
21. Quater, P. B., Grimaccia, F., Leva, S., Mussetta, M., & Aghaei, M. (2014). Light Unmanned Aerial Vehicles (UAVs) for cooperative inspection of PV plants. *IEEE Journal of Photovoltaics*, 4(4), 1107–1113.
22. Koch, S., Weber, T., Sobottka, C., Fladung, A., Clemens, P., & Berghold, J. (2016, June). Outdoor electroluminescence imaging of crystalline photovoltaic modules: Comparative study between manual ground-level inspections and drone-based aerial surveys. In *32nd European Photovoltaic Solar Energy Conference and Exhibition* (pp. 1736–1740).
23. Buerhop-Lutz, C., & Scheuerpflug, H. (2015). Inspecting PV-plants using aerial, drone-mounted infrared thermography system. *3rd Southern African Solar Energy Conference, South Africa*, 11–13 May, 2015.
24. Aghaei, M., Gandelli, A., Grimaccia, F., Leva, S., & Zich, R. E. (2015, June). IR real-time analyses for PV system monitoring by digital image processing techniques. In *2015 International Conference on Event-Based Control, Communication, and Signal Processing (ebccsp)* (pp. 1–6). IEEE.
25. Ji, D., Zhang, C., Lv, M., Ma, Y., & Guan, N. (2017). Photovoltaic array fault detection by automatic reconfiguration. *Energies*, 10(5), 699.
26. Kuitche, J. M., Pan, R., & Tamizh Mani, G. (2014). Investigation of dominant failure mode (s) for field-aged crystalline silicon PV modules under desert climatic conditions. *IEEE Journal of Photovoltaics*, 4(3), 814–826.
27. Cristaldi, L., Khalil, M., & Soulatiantork, P. (2017). A root cause analysis and a risk evaluation of PV balance of system failures. *Acta Imeko*, 6, 113–120.
28. Colli, A. (2015). Failure mode and effect analysis for photovoltaic systems. *Renewable and Sustainable Energy Reviews*, 50, 804–809.
29. Villarini, M., Cesarotti, V., Alfonsi, L., & Introna, V. (2017). Optimization of photovoltaic maintenance plan by means of a FMEA approach based on real data. *Energy Conversion and Management*, 152, 1–12.

30. Shrestha, S. M., Mallineni, J. K., Yedidi, K. R., Knisely, B., Tatapudi, S., Kuitche, J., & Tamizh Mani, G. (2014). Determination of dominant failure modes using FMECA on the field deployed c-Si modules under hot-dry desert climate. *IEEE Journal of Photovoltaics*, 5(1), 174–182.
31. Catelani, M., Ciani, L., Cristaldi, L., Faifer, M., & Lazzaroni, M. (2013). Electrical performances optimization of Photovoltaic Modules with FMECA approach. *Measurement*, 46(10), 3898–3909.
32. Gallardo-Saavedra, S., Hernández-Callejo, L., & Duque-Pérez, O. (2019). Quantitative failure rates and modes analysis in photovoltaic plants. *Energy*, 183, 825–836.
33. Marquez, R., & Coimbra, C. F. (2011). Forecasting of global and direct solar irradiance using stochastic learning methods, ground experiments and the NWS database. *Solar Energy*, 85(5), 746–756.
34. Wilson, I. D., Jones, A. J., Jenkins, D. H., & Ware, J. A. (2004). Predicting housing value: genetic algorithm attribute selection and dependence modelling utilising the Gamma test. In *Applications of Artificial Intelligence in Finance and Economics* (pp. 1–27). Emerald Group Publishing Limited.
35. Zamora, R. J., Dutton, E. G., Trainer, M., McKeen, S. A., Wilczak, J. M., & Hou, Y. T. (2005). The accuracy of solar irradiance calculations used in mesoscale numerical weather prediction. *Monthly Weather Review*, 133(4), 783–792.
36. Devabhaktuni, V., Alam, M., Depuru, S. S. S. R., Green II, R. C., Nims, D., & Near, C. (2013). Solar energy: Trends and enabling technologies. *Renewable and Sustainable Energy Reviews*, 19, 555–564.
37. Rivai, A., & Rahim, N. A. (2013, November). A low-cost photovoltaic (PV) array monitoring system. In *2013 IEEE Conference on Clean Energy and Technology (CEAT)* (pp. 169–174). IEEE.
38. Chouder, A., Silvestre, S., Taghezouit, B., & Karatepe, E. (2013). Monitoring, modelling and simulation of PV systems using LabVIEW. *Solar Energy*, 91, 337–349.
39. Silvestre, S., Chouder, A., & Karatepe, E. (2013). Automatic fault detection in grid connected PV systems. *Solar Energy*, 94, 119–127.
40. Prieto, M. J., Pernía, A. M., Nuño, F., Díaz, J., & Villegas, P. J. (2014). Development of a wireless sensor network for individual monitoring of panels in a photovoltaic plant. *Sensors*, 14(2), 2379–2396.
41. Adhya, S., Saha, D., Das, A., Jana, J., & Saha, H. (2016, January). An IoT based smart solar photovoltaic remote monitoring and control unit. In *2016 2nd international conference on control, instrumentation, energy & communication (CIEC)* (pp. 432–436). IEEE.
42. Shariff, F., Abd Rahim, N., & Ping, H. W. (2013, November). Photovoltaic remote monitoring system based on GSM. In *2013 IEEE Conference on Clean Energy and Technology (CEAT)* (pp. 379–383). IEEE.
43. De Benedetti, M., Leonardi, F., Messina, F., Santoro, C., & Vasilakos, A. (2018). Anomaly detection and predictive maintenance for photovoltaic systems. *Neurocomputing*, 310, 59–68.
44. Montgomery, D. C., & Runger, G. C. (2007). *Applied statistics and probability for engineers, (With CD)*. New York: John Wiley & Sons.

6 Optimization Modeling Techniques for Energy Forecasting and Condition-Based Maintenance in PV Plants

K. Lakshmi, G. Sophia Jasmine, and D. Magdalin Mary
Sri Krishna College of Technology, Coimbatore, India

CONTENTS

DOI: 10.1201/9781003202288-6

LEARNING OUTCOMES

i. To review the current scenario and future scope of solar energy
ii. To learn about the types of solar forecasting techniques and various optimization methods used in solar energy forecasting
iii. To study condition-based maintenance in PV plants

6.1 SCENARIO OF SOLAR ENERGY

6.1.1 Present Scenario of Solar Energy in India and Future Scope

India has been on an aggressive path toward becoming a developed country with an ambitious target of becoming a US$5 trillion economy by 2025. The Indian economy consists of agriculture, fishing, and mining & quarrying (20%); manufacturing, electricity, water supply, other utility services, and construction (25%); and the largest services sector of administration, defense, finance, trade, hospitality, and IT services (55%) (Figures 6.1, 6.2).

The services sector's share in the gross domestic product (GDP) has almost doubled from 30% in 1950 to 55% presently. The power consumption of India as a whole has risen from65 MW peak demand in 1995 to 185 MW in 2019 (Rachit Srivastava 2016). To continue fueling economic expansion, India must possess a robust and ever-increasing power generation capacity. India has managed to increase the power generation capacity from 179 GW in 2009 to 379 GW in 2021.

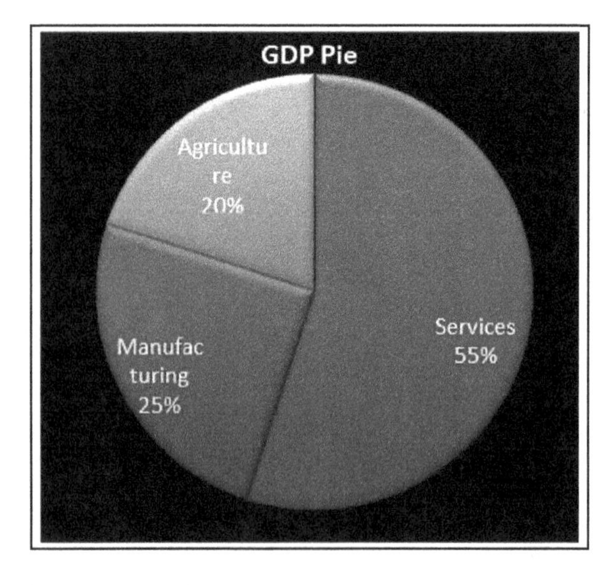

FIGURE 6.1 Various sectors of Indian economy.

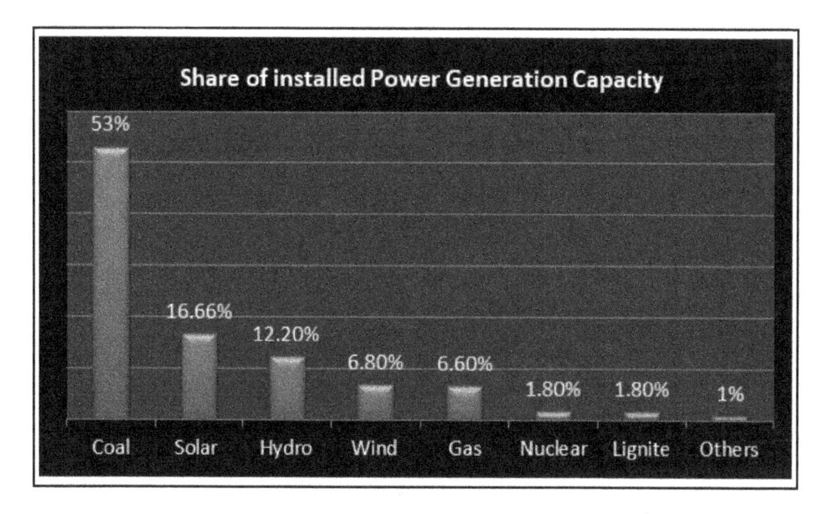

FIGURE 6.2 Comparison of power generation capacity.

The 379 GW power generation capacity has two major constituents: nonrenewable energy sources, accounting for 60% of the total capacity; and renewable energy sources, accounting for the remaining 40%. The share of energy generation fuels includes coal (53%), natural gas (6.6%), lignite and diesel (1.8%), hydro (12.2%), nuclear (1.8%), and other renewable sources (24.5%). With the focus on reducing the use of fossil fuels that pollute the environment and moving toward "greener" solutions, the use of renewable sources is on the rise. India experiences an average of 300 days of bright sunlight a year. The average solar energy generation capacity of India is estimated to be around 420 trillion KWh every month. The national institute of solar energy has estimated that by utilizing just 3% of the wasteland, India has the potential of generating up to 780 GW of electricity from solar—almost double the current installed energy capacity.

6.1.2 National Solar Mission

The Indian government formulated the national solar mission as one of the many policies contributing to the national action plan for climate change. The initial target was 20 GW by 2022. India had solar energy generation capacity of 2.65 GW in May 2014 and increased it to 20 GW in January 2021. In 2015, the government revised the target to 100 GW by 2022, and as of March 2021, the achieved capacity was 40 GW, with further 37.10 GW under installation and another 21.21 GW tendered for bidding. Once this capacity comes online, it will bring the total solar energy generation capability of the country to 94.6 GW. As per the Ministry of Statistics and Program Implementation (MOSPI) website, solar energy constituted 68% of total renewable resources available in India as of 2018. The country has established over 42 solar parks, and the Bhadla solar park in Rajasthan is the largest in the world in terms of power generated and second largest in physical size (Figure 6.3).

FIGURE 6.3 Second largest power generation.

In his speech at the 15th G20 summit, Prime Minister Narendra Modi has declared that India has set a target of 450 GW of solar energy generation capacity by 2030 (Sign and Singh 2016).The union government has been aggressively implementing 2 million standalone solar pumps under the scheme Pradhan Mantri Kisan Urja Suraksha evam Utthaan Mahabhiyan (PM-KUSUM). This scheme allows farmers to install solar-powered pumps for irrigation. As of October 2020, 213,000 solar lamps and 1.5 MW of solar power packs have been installed under the Off-Grid Solar PV Applications Programme Phase III, and 80,000 solar street lights have been installed out of the total target of 150,000 units under the Atal Jyoti Yojana (AJAY) Phase-II (Naveen Shankarapp et al. 2017). To date, 4.4 GW of rooftop solar power generators have been installed, against the target of 40 GW by 2022.

6.1.3 Tariff Trend Comparison of Various Fuels Used for Power Generation

The continuous fall in the cost of power generation from solar energy is the main force driving the policy makers to shift toward more renewable sources. As per International Energy Agency, power generation using solar energy has become cheaper than using coal in 2020 worldwide (Figures 6.4, 6.5).

6.1.4 Cost of Installation of a Rooftop Solar Power Plant: Financial Savings and Reduction in CO_2 Emissions

Figure 6.6 presents an estimate of installing a rooftop solar power plant in Tamil Nadu over an area of 1,300 square feet. A return on investment (ROI) would be achieved within 5 years, and the savings would extend for another 20 years. It has been reported that about 23 domestic solar PV modules are installed in India as of March 2021 (Ken Brook 2021) (Figure 6.6).

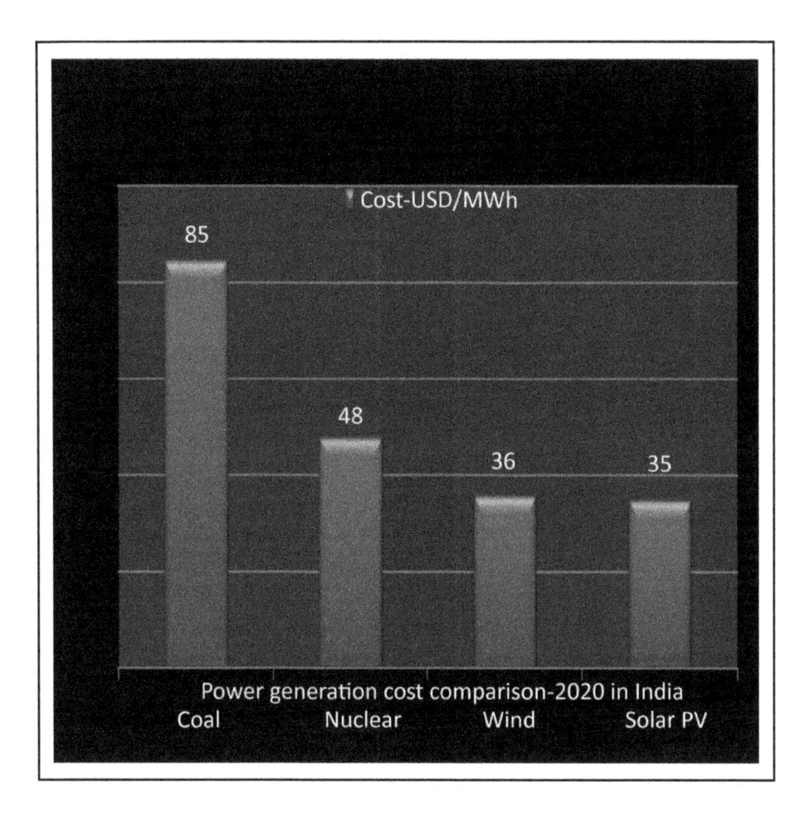

FIGURE 6.4 Power generation cost comparison.

6.1.5 Grid-Connected Solar PV System Block Diagram and Cost Requirement

A grid-connected solar PV system consists of a PV array, maximum power point tracker (MPPT), charge controller, inverter, battery storage, controller grid interface, and grid interface for different kinds of works. MPPT is an electronic DC-to-DC converter that optimizes the value between the solar array and the battery storage (Bagalini et al. 2019). An inverter is used to convert DC current into AC current. The converted output is the delivered to a utility grid (Figure 6.7).

6.1.6 Future of Solar Power Generation

To date, India has 37 GW of solar energy generation capacity installed. The target set by the government is 100 GW of solar plants and a total 175 GW target for renewable sources. Smaller targets compared to solar are set for wind energy at 60 GW, hydro at 10 GW, and biomass at 5 GW. While solar PV plants installations have kept pace with the designated targets, rooftop solar plants are yet to pick up pace in order for India to achieve this ambitious target.

Rapid economic development, sustained annual GDP growth of 8%, limited supply of coal, its low level of technological improvements and environmentally hazardous nature, smaller domestic reserves of hydrocarbons and dependency on uncertain

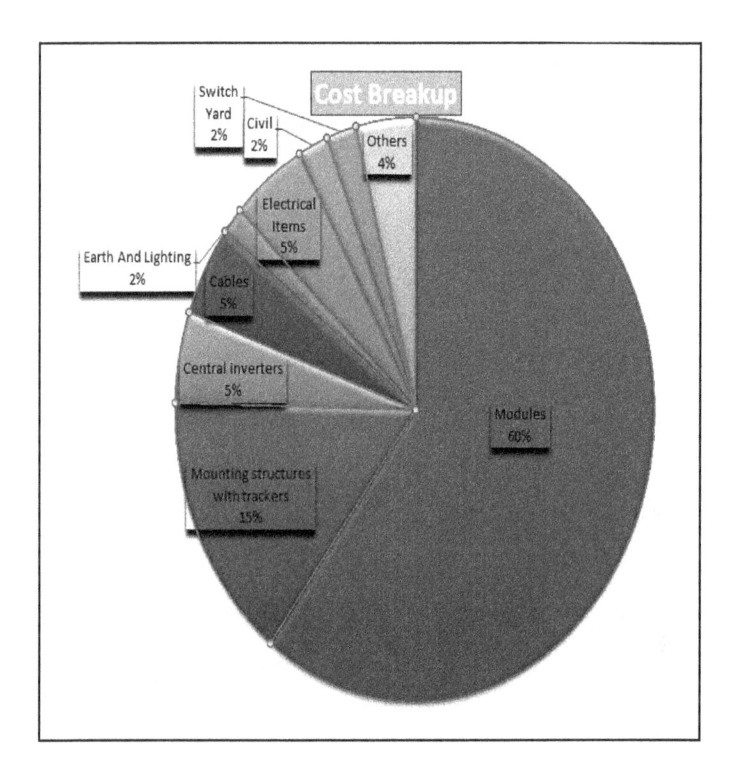

FIGURE 6.5 Cost breakdown in various sectors.

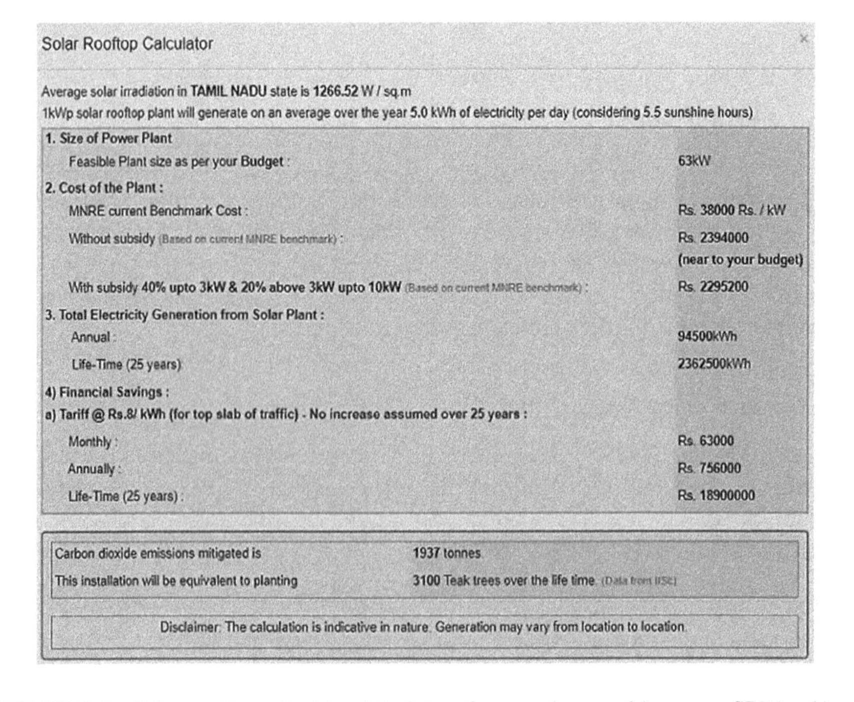

Solar Rooftop Calculator

Average solar irradiation in **TAMIL NADU** state is **1266.52** W / sq.m

1kWp solar rooftop plant will generate on an average over the year **5.0** kWh of electricity per day (considering 5.5 sunshine hours)

1. Size of Power Plant	
Feasible Plant size as per your **Budget** :	63kW
2. Cost of the Plant :	
MNRE current Benchmark Cost :	Rs. 38000 Rs. / kW
Without subsidy (Based on current MNRE benchmark) :	Rs. 2394000 **(near to your budget)**
With subsidy 40% upto 3kW & 20% above 3kW upto 10kW (Based on current MNRE benchmark) :	Rs. 2295200
3. Total Electricity Generation from Solar Plant :	
Annual :	94500kWh
Life-Time (25 years) :	2362500kWh
4) Financial Savings :	
a) Tariff @ Rs.8/ kWh (for top slab of traffic) - No increase assumed over 25 years :	
Monthly :	Rs. 63000
Annually :	Rs. 756000
Life-Time (25 years) :	Rs. 18900000

Carbon dioxide emissions mitigated is	**1937** tonnes
This installation will be equivalent to planting	**3100** Teak trees over the life time. (Data from IISc)

Disclaimer: The calculation is indicative in nature. Generation may vary from location to location.

FIGURE 6.6 Solar rooftop calculator (Ministry of new and renewable energy SPIN n.d.).

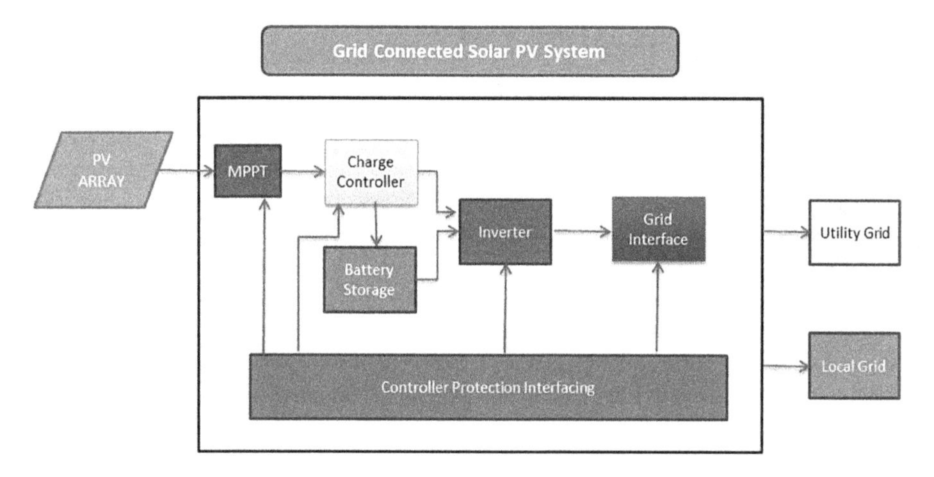

FIGURE 6.7 Grid-connected solar PV system.

foreign supply, and rising costs of fossil fuels directly impact the government to focus more on renewable sources. As per the commitments to the Paris Climate Accord, India has set an ambitious target of reducing its dependency on coal by 50% in 2040.

For all the strategies, regulations, inventions, and corporate requirements, the integration of solar energy–generated electricity into the power grids is facing problems stemming mainly from infrastructure incompatibilities and the need to predict weather conditions across long periods of time with a high degree of accuracy. Effective forecasting techniques will help increase the value of solar power plants, and providing optimal energy transition between a conventional grid and a PV array will help ensure continuity of supply (Lotfi et al. 2020).

6.2 NEED FOR ENERGY FORECASTING IN PV PLANTS

Solar energy forecasting plays a major role in improving performance and interoperability of a solar power plant. It requires the knowledge of the sun's movement across the sky, atmospheric conditions, sunlight scattering processes, and performance characteristics of the PV plant. Some auxiliary inputs, such as temperature variations across a PV panel, are also useful. Photovoltaic systems convert solar energy into electricity using semiconducting materials (Lotfi et al. 2020). Accurate analysis of solar output depends on data availability and methods used. These include statistical learning methods (stochastic), sensing methods (local and remote sensing), and collaborative techniques (hybrid) that combine different kinds of forecasts.

This chapter suggests the best approach for constructing a PV power forecasting method, which includes the analysis of the time perspectives between the present and future times (APV Appa Rao et al. 2018). Forecasting is done across multiple time horizons, from very short ones (measured in hours or even minutes) that are essential for operation of utility centers and micro-grids, to medium ones (several days to several months) essential for predictive maintenance, to long, three-dimensional (time and space) horizons used in regional forecasting. This chapter explores the methods used for all time horizons of forecasting.

6.3 ENERGY FORECASTING TYPES

6.3.1 Significance of Energy Forecasting

Solar energy forecasting is linked to analyzing the weather conditions. It is split into two categories: (1) studying the patterns of climatological variables and (2) estimating the amount of energy produced under varying weather conditions. Dealing with these issues is related to the three-dimensional and progressive scales, which yields to different approaches.

Solar radiation is powerful enough to affect air temperature, moisture in the soil, water vaporization, precipitation levels, and other photochemical reactions by effecting physical and biological changes on the surface of the Earth. Solar radiation is the driving force of efficient life cycle management of both plants and animals and a very important feature in understanding forestry sciences. The amount of radiation reaching the ground depends on the terrain of the particular region, its location on Earth, and its animal and plant life. Detailed knowledge of these and other details should allow for a more precise forecasting of the amount of solar radiation to reach the location where a PV power plant is planned. The technologies used in preparing these models include Geographical Information Systems (GIS), artificial intelligence, numerical weather forecast (NWP). Their effectiveness depends on experiential statistics, which have a superior precision compared to an extra method. Therefore, while an adequate footage spatial thickness is accessible, a disturbance method is preferred. Conventionally, solar radiation has not been as important a variable as temperature or rainfall. There are several experimental systems that record solar radiation, and the interpretation has been converted into an appropriate technique for a solar radiation evaluation standard. Radiometric stations are more prevalent in farmland or around basin and plains, while the mountain regions require enough footage thickness. Hence, a particular interruption method, a more affordable procedure to measure the tall spatial and unpredictable solar radiation in hilly areas in an affordable manner, is employed. Still there are few, if any, easy-to-use methods of evaluation in mountainous regions. As a result, it is hard to conduct a precise solar radiation analysis in hilly areas.

It is necessary to install radiometric sensors in order to illustrate the changes in solar radiation with respect to the supplementary ecological variables such as temperature or precipitation, because this particular parameter is extremely sensitive to these ecological features. Terrain surface usually confounds the conventional interruption method, mainly because measurement stations are sparse in mountainous areas. Geostatistics help in a stochastic move to resolve the spatial forecast difficulty and reduce the dependence on deterministic models (Figure 6.8).

6.3.2 Short-Term Solar Power Forecasting

Short-term forecasting provides predictions up to seven days ahead, which is helpful in decision-making concerning grid operations. Meteorological resources are estimated with a different temporal and spatial resolution in short-term forecasting. In this sense, most of the approaches make use of different numerical weather prediction models (NWP) that provide an initial estimation of weather variables. While

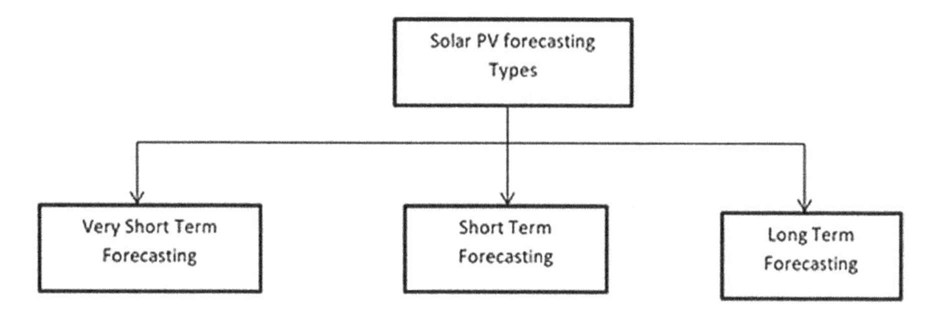

FIGURE 6.8 Types of energy forecasting.

there are several models to choose from, Global Forecast System (GFS) or the one provided by the European Center for Medium Range Weather Forecasting (ECMWF) is the preferred one. These two models provide meteorological forecasts all over the world and are more sophisticated than the alternatives. In order to further increase spatial and temporal resolution of these models, mesoscale models are developed. HIRLAM, WRF, or MM5 are the most widely used models worldwide. Application of these models requires extensive expertise due to a wide variety of parameters available for configuration (Samuel et al. 2014). Data assimilation is used alongside these models in order to produce more realistic simulations. Once the model outputs are ready, ensemble techniques, like the one proposed by Bacher et al. (2009), that mix outputs of different models and provide a probabilistic view of the accuracy of the output, are used (Utpal Kumar Das et al. 2018a). This will provide a better estimate of all variables and reduce the degree of uncertainty.

6.3.3 Long-Term Solar Power Forecasting

Long-term forecasting usually refers to forecasting of resources available monthly or annually. This is useful for large-scale energy producers in negotiating contracts with financial entities or utilities that distribute the generated energy. Hence, most of these models are run with mesoscale models fed with reanalysis data as input and the output is postprocessed with statistical approaches of the measured data.

6.3.4 Energetic Models

Any output from any model must be converted to the electric energy that a particular solar PV plant will produce. This is usually achieved by using statistical approaches that try to correlate the amount of available resource with the metered power output. The main advantage of these methods is that the meteorological prediction error, which is the main component of the global error, might be reduced considering the uncertainty of the prediction. As it was mentioned before and detailed in Heinemann et al., these statistical approaches consist of ARMA models, neural networks, support vector machines, etc. On the other hand, there also exist theoretical models that describe how a power plant converts the meteorological resource into electric energy, as described in Alonso et al. The main advantage of this type of models is that when they are fitted, they are highly accurate, although they are too sensitive to the meteorological prediction error, which is usually amplified by these models. Hybrid models are a combination of both Heinemann el al.'s and Alonso et al.'s models and represent a promising approach that can outperform each of them individually.

6.4 TECHNIQUES USED FOR SOLAR ENERGY FORECASTING

Solar energy is eco-friendly and affords electrical power to locations where conventional power plants cannot be built. It has been reported that the intensity of solar radiation on the Earth's surface is 1,367 W/m^2 and the absorption of solar energy is 1.8×10^{11} MW (Shah et al. 2015). It is inevitable that solar power generation is completely weather dependent and that power production is possible only during daytime. The foremost challenge in PV power production is the forecasting of solar energy. Forecasting is a vital and cost-effective means for integrating variable renewable energy (VRE) resources like solar into power systems. Various approaches such as physical, statistical, and artificial intelligence (optimization and machine learning algorithms)-based forecasting methods have been exploited for better PV output. These approaches along with meteorological statistics have the potential to overcome the blockades related to weather conditions.

The maximum power output from the PV cell is demonstrated by Pandiarajan and Muthu (2011),

$$\mathrm{Pmax} = \eta_{\mathrm{solar}}\, A_{\mathrm{arrary}}\, I_{\mathrm{solar}}\left[1 - 0.05\left(t_{\mathrm{atm}} - 25\right)\right]$$

where η_{solar} is the PV array conversion efficiency, A_{arrary} is the solar panel area in m^2, I_{solar} is the radiance of the sun in kW/m^2, and t_{atm} is the temperature of the atmosphere in degree Celsius. Based on such modeling, various techniques were proposed by researchers around the globe to forecast the solar energy for its better utilization and maximum production of electrical power.

6.4.1 OPTIMIZATION MODELING TECHNIQUES

The choice of input parameter is the most significant criterion for better forecasting output. Hence optimization tools play a major role in producing optimized input values. Radial Basis Function Neural network (RBFN) model of online short-term forecasting has been reported by the researcher. The RBFN forecasting model as shown in Figure 6.9 predicts the solar system output 24 hours ahead (Chen et al. 2011).

Multilayer perceptron neural network (MLPNN), as given in Figure 6.10, is considered as a yardstick by many researchers for predicting the solar energy (Bui et al. 2016). Recurrent Neural Network (RNN) is an extended type of ANN that can learn and advance various multifaceted and composite relationships between computational structures. RNN depends on time sequence statistics by a feedback system to receive the preceding period phase values, signifying progressive dynamic features.

For accurate forecasting, it is often desirable to predict the upper and lower values of solar radiation, called prediction interval, which can be accomplished using the neural network (Galvan et al. 2017, Abedinia et al. 2017).

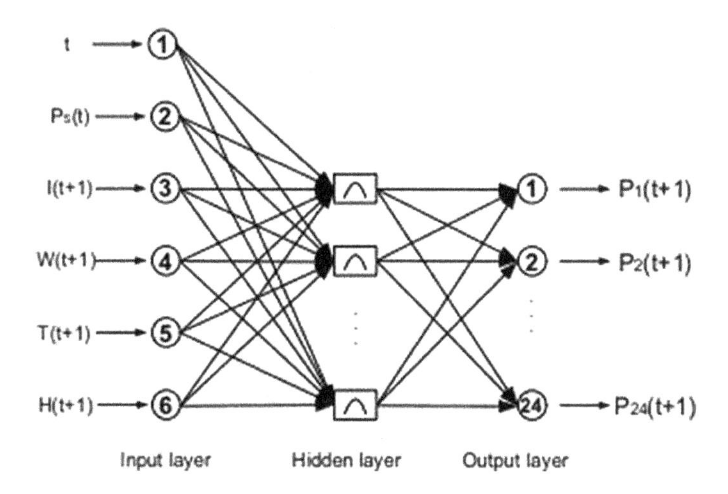

FIGURE 6.9 RBFN model for forecasting solar energy 24 hours ahead (Chen et al. 2011).

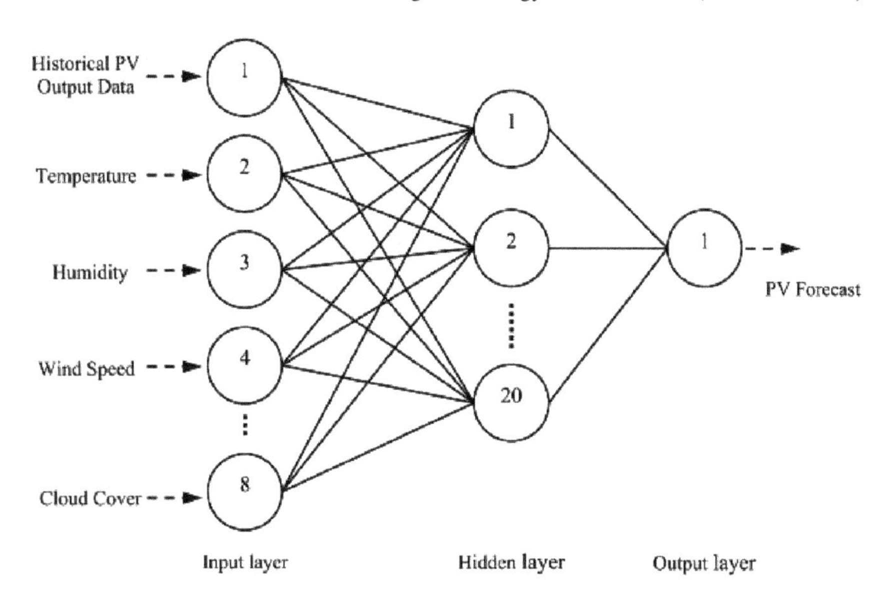

FIGURE 6.10 Multilayered perceptron neural network (Raza et al. 2016).

Fuzzy logic, ANFIS, and ANN are used extensively in solar energy forecasting. A relative performance (Perveen et al. 2019) has been investigated for four different weather conditions at five different locations in India. Among various techniques, ANN and SVM (Support Vector Machine) (Das et al. 2018b) have proven their efficacy under varying weather conditions. The accuracy of the forecasted data will be high when the preprocessed data are used as the input. Also, it is expected to have a high correlation factor between the input and output. The accuracy of the forecasted values depends not only on the input used but also on the forecast horizon. Genetic Algorithm–based Neural Network (Tao and Chen 2014) has been used to forecast the solar energy, and the sample data obtained are from Henan University of Science and Technology,

China. Genetic Algorithm (GA) is an effective method (Liu et al. 2018) in hydro-PV integrated system and achieved good complementarity between hydropower and PV systems. GA combined with ANFIS (Yadav et al. 2019) provides a flexible technique for short-term solar PV power forecasting with an error of approximately 12%. Hybrid PSO-ANFIS (Semero et al. 2018) technique finds its application in short-term PV forecasting in microgrids by providing daily average forecasted error of less than 7%.

6.4.2 Machine Learning Techniques

Machine learning (ML) is the science of making computers learn like humans do. It can be used in various domains, as it has the advantage of producing reliable solution that cannot be achieved using any other explicit algorithms. The most important part of ML technique is classification and data mining. Based on the 10 months of solar radiation data, ML is applied to predict solar radiation (Sharma et al. 2011) considering other inputs such as temperature, wind speeds, humidity, etc. ML is applied in monitoring the PV panels as well (Haba 2019) (Figure 6.11).

ML finds a relation between input and output. Efficiency of ML output depends on the right data. Learning can be done using linear regression, generalized linear models, nonlinear regression, support vector regression, decision tree learning, and hierarchical clustering.

6.5 CONDITION-BASED MAINTENANCE IN PV PLANTS

Condition-based maintenance is a preservation approach that monitors the real ailment of a device or a system to resolve it by adopting any preservation method. Often the maintenance is scheduled in advance, as in power stations. As the solar panels are consistently exposed to atmosphere, condition-based maintenance needs to be followed rather than scheduled maintenance, and thereby it is possible to monitor and find out the upcoming failure in the panel if any. Data can be collected for such maintenance by two means: (1) sensor reading can be collected at regular intervals; or (2) sensor readings can be collected continuously. The development of smart sensors, digital devices, Internet of Things, etc. has led to the creation of effective and consistent technologies for enhancing the maintenance plans. An organized condition-based maintenance plan is presented in Figure 6.12.

6.5.1 Benefits of Condition-Based Maintenance

1. Number of accidental failures is reduced.
2. Reliable operation is achieved.
3. Maintenance work is possible during non-peak hours.
4. Lifetime of the PV panel is increased.
5. Inventory cost is reduced.
6. As maintenance is done only when needed, time spent for repairing is less compared with scheduled maintenance.

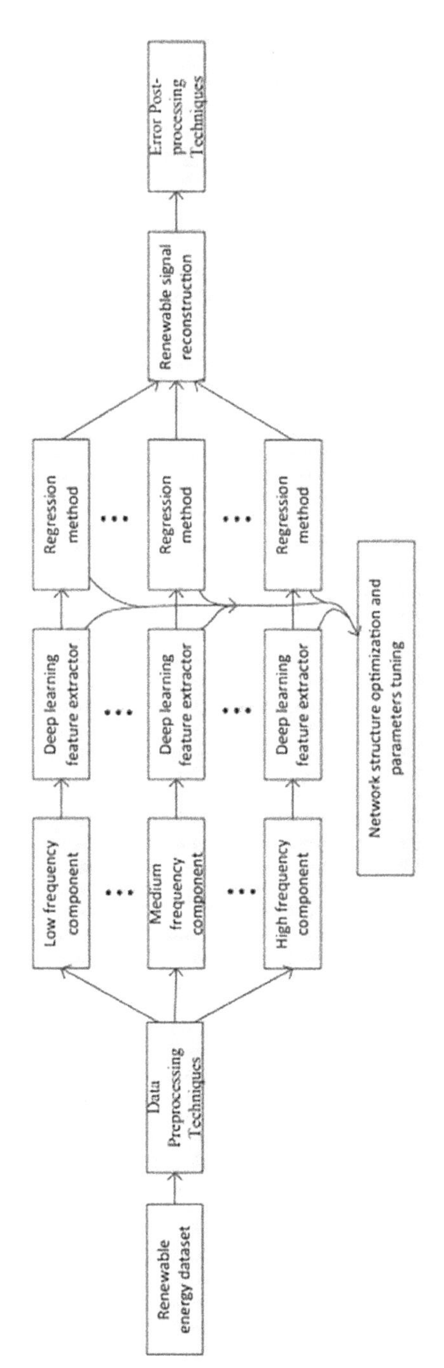

FIGURE 6.11 Schematic of general renewable energy forecasting (Wang et al. 2019).

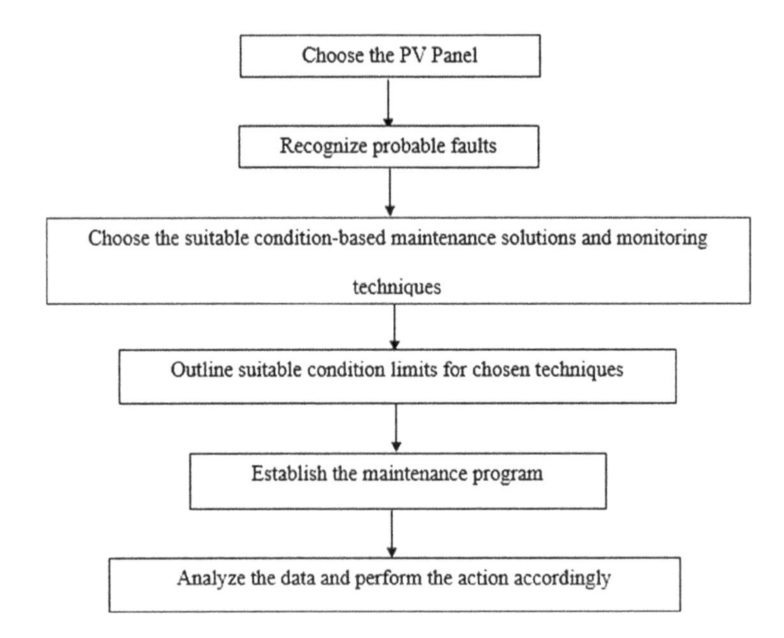

FIGURE 6.12 Organization of condition-based maintenance program.

CASE STUDY

As solar power production is solely weather dependent, forecasting of solar energy is a vital means in solar PV plant to enhance power production. **Subhra Das (2020)** discusses a case study on the short-term forecasting of 89.6 kWp solar PV power plant. The solar panels were installed at the rooftop of a block at Amity University, Haryana. In this work, the technique used to forecast solar power is autoregressive integrated moving average (ARIMA) model. Root mean square error has been estimated and found to be considerably smaller when compared with analytical models.

Jesus Ferrero Bermejo et al. (2019) detail a case study on designing a predictive analytics for the improvement of condition-based maintenance (CBM) strategies and energy production forecasting. Intelligent data analysis (IDA) is considered for building this system. Solar radiance, temperature, and humidity are measured as inputs using sensors that continuously feed data to SCADA systems. These inputs are analyzed using data mining (DM) methods. These inputs are fed to various DM techniques such as artificial neural networks (ANN), support vector machine (SVM), decision trees, etc. The results of each technique are then compared and the best model is selected to design a predictive analysis model.

Sameer Al-Dahidi et al. (2019) underscores the importance of identifying a forecasting model for balancing the variable of and increasing power inputs from plants with intermittent energy sources. Extreme learning machines (ELM) module of simple, fast computational and good generalization capacity is selected. The methodology used was taking input of around 45 weather variables. The RMSE, MAE, and WMAE values of the experimental models are compared with BP ANN model. Based on the comparison, ELM model is slightly more accurate than BP ANN model in solar photovoltaic power predictions.

Ahmed et al. (2020) detail different factors affecting PVPF accuracy making such prediction a sophisticated process. It depends on factors such as forecasting horizons, forecast model inputs, and performance estimation. To achieve better precision, correlational analysis optimization and uncertainty estimation of PVPF models need to be carried out. Another factor in cloud status–based PVPF modeling is the use of ground-based sky images, which have higher resolution than satellite imagery. The authors reclassified the standard 33 meteorological weather categories into 10 weather classes by compiling several single weather types to constitute a single new weather type. Solar radiance, clouds, temperature, wind speed and cell temperature, and air pressure are some of the input parameters whose data are recorded over a span of two years on an hourly basis. The literature describes many optimization techniques for inputs of PV output models: PSO, grid search, fruit fly optimization, firefly, ant colony optimization, and chaotic ant swarm optimization. Forecasting techniques are classified as persistence forecast, physical model, time series–based forecasting models, exponential smoothing, the autoregressive moving average model (ARMA), artificial numeral network (ANN), online PVPF, sequential extreme learning machine (OS-ELM), and ensemble of models approach to PVPF.

6.6 SUMMARY

Solar energy is a great boon for tropical countries where sunlight is plentiful. As this resource is weather dependent, it must be forecasted for better efficiency. Optimization techniques afford a solution for predicting the solar energy. The chapter discusses the need for solar energy forecasting, its types, optimization techniques used for solar forecasting, and condition-based maintenance in PV plants. Various optimization techniques and their applications in energy forecasting have been discussed. PSO finds its extensive application in PV solar energy forecasting amid many other optimization algorithms.

REFERENCES

Oveis Abedinia, Nima Amjady and Noradin Ghadimi, Solar energy forecasting based on hybrid neural network and improved metaheuristic algorithm, *Computational Intelligence* (2017), doi:10.1111/coin.12145.

V. Ahmed, Y. Sreeram and M.D. Arif Mishra, A review and evaluation of the state-of-the-art in PV solar power forecasting: Techniques and optimization, *Renewable and Sustainable Energy Reviews* (2020), Vol. 124, pp. 777–780.

Sameer Al-Dahidi, Osama Ayadi, Jehad Adeeb, Mohammad Alrbai and Bashar R. Qawasmeh, Extreme learning machines for solar photovoltaic power predictions, *Energies* (2019), doi:10.3390/en11102725.

V. Bagalini, B.Y. Zhao, R.Z. Wang and U. Desideri, Solar PV-battery-electric grid-based energy system for residential applications: System configuration and viability, *Research Article Open Access* (2019), Vol. 2019, Article ID 3838603, doi:10.34133/2019/3838603.

Manoja Kumar Beher, Irani Majumder and Niranjan Nayaka, Solar photovoltaic power forecasting using optimized modified extreme learning machine technique, *Engineering Science and Technology, International Journal* (2018), Vol. 21, No. 3, pp. 428–438.

Jesus Ferrero Bermejo, Juan Francisco Gomez Fernandez, Rafael Pino, Adolfo Crespo Marquez and Antonio Jesus Guillen Lopez, Review and comparison of intelligent optimization modelling techniques for energy forecasting and condition-based maintenance in PV plants, *Energies* (2019), Vol. 12, doi:10.3390/en12214163.

K.-T.T. Bui, D. Tien Bui, J. Zou, C. Van Doan and I. Revhaug, A novel hybrid artificial intelligent approach based on neural fuzzy inference model and particle swarm optimization for horizontal displacement modeling of hydropower dam. *Neural Computing & Applications* (2016), Vol. 29, No. 12, pp. 495–506.

Changsong Chen, Shanxu Duan, Tao Cai and Bangyin Liu, Online 24-h solar power forecasting based on weather type classification using artificial neural network, *Solar Energy* (2011), Vol. 85, No. 11, pp. 2856–2870.

Subhra Das, Short term forecasting of solar radiation and power output of 89.6 kWp solar PV power plant, *Materials Today: Proceedings* (2020), doi:10.1016/j.matpr.2020.08.449

Utpal Kumar Das, Kok Soon Tey, Mehdi Seyedmahmoudian, Saad Mekhilef, Moh Yamani Idna Idris, Willem Van Deventer, Bend Horan and Alex Stojcevski, Forecasting of photovoltaic power generation and model optimization: A review, *Renewable and Sustainable Energy Reviews* (2018a), Vol. 81, pp. 912–928.

Utpal Kumar Das, Kok Soon Teya, Mehdi Seyedmahmoudian, Saad Mekhilef, Moh Yamani Idna Idris, Willem Van Deventer, Bend Horan and Alex Stojcevski, Forecasting of photovoltaic power generation and model optimization: A review, *Renewable and Sustainable Energy Reviews* (2018b), Vol. 18, pp. 912–928.

Abdelhakim El Hendouzi and Abdennaser Bourouhou, Solar photovoltaic power forecasting, *Hindawi Journal of Electrical and Computer Engineering* (2020), Vol. 21, Article ID 8819925, doi:10.1155/2020/8819925.

Ines M. Galvan, Jose M. Valls, Alejandro Cervantes and Ricardo Aler, Multi-objective evolutionary optimization of prediction intervals for solar energy forecasting with neural networks, *Journal on Information Sciences* (2017), Vol. 148, pp. 363–382.

Cristian-Gvozo Haba, Monitoring solar panels using machine learning techniques. In *International Conference on Modern Power Systems (MPS)* (2019), doi:10.1109/MPS.2019.8759651.

Luyao Liu, Qie Sun, Yu Wang, Yiling Liu and Ronald Wennersten, Research on short-term optimization for integrated hydro-PV power system based on genetic algorithm, *Energies* (2018), Vol. 152, pp. 1097–1102.

Mohamed Lotfi, Mohammad Javadi, Gerardo J. Osório, Cláudio Monteiro and P.S. Catalao João, A novel ensemble algorithm for solar power forecasting based on kernel density estimation, *Energies* (2020), Vol. 13, No. 216, doi:10.3390/en13010216.

Ministry of New and Renewable Energy SPIN, An online application for Solar Photovoltaic Installation: (Grid-Connected Rooftop), (n.d.) https://solarrooftop.gov.in/rooftopcalculator

N. Pandiarajan and R. Muthu, Mathematical modeling of photovoltaic module with Simulink. In *1st International Conference on Electrical Energy Systems (ICEES)* (2011), pp. 258–263.

Gulnar Perveen, Mohammad Rizwan and Nidhi Goel, Comparison of intelligent modelling techniques for forecasting solar energy and its application in solar PV based energy system, *IET Energy Systems Integration* (2019), Vol. 1, No. 1, pp. 34–51.

APV Appa Rao, J. Rama Mohan, L. Malleswara Rao, K. Ram Narayana, C.H. Sundar Singh and P. Ramakrishna Rao, Solar energy in India – present and future, *International Journal of Engineering Science Invention (IJESI)* (2018), ISSN (Online): 2319-6734, ISSN (Print): 2319-6726.

M.Q. Raza, M. Nadarajah and C. Ekanayake, On recent advances in PV output power forecast, *Solar Energy* (2016), Vol. 136, pp. 125–144.

Isaac A. Samuel, F. Felly-Njoku Chihurumanya, A. Adewale Adeyinka and Ayokunle A. Awelewa, Medium-term load forecasting of covenant university using the regression analysis methods, *Journal of Energy Technologies* (2014), Vol. 4, No. 4, pp. 2224–2232.

Yordanos Kassa Semero, Dehua Zheng and Jianhua Zhang, A PSO-ANFIS based hybrid approach for short term PV power prediction in microgrids, *Electric Power Components and Systems* (2018), doi:10.1080/15325008.2018.1433733.

A. Shah, H. Yokoyama and N. Kakimoto, High-precision forecasting model of solar irradiance based on grid point value data analysis for an efficient photovoltaic system. *IEEE Trans Sustain Energy* (2015), Vol. 6, No. 2, pp. 474–481.

Naveen Shankarapp, Mufassireen Ahmed, N. Shashikiran and H. Naganagouda, Solar photovoltaic systems – applications & configurations, *International Research Journal of Engineering and Technology (IRJET)* (2017),Vol. 4, No. 8, doi:10.1109/SmartGridComm.2011.6102379.

Navin Sharma, Pranshu Sharma, David Irwin and Prashant Shenoy, Predicting Solar Generation from weather forecasts using machine learning, *IEEE Smart Grid Communications* (2011), doi:10.1109/smartgridcomm.2011.6102379.

Bharat Raj Sign and Onkar Singh, Future scope of India, *Journal of Physical Sciences Engineering and Technology* (2016), doi:10.18090/samriddhi.v8i1.11408.

Ken Brook Solar, MW Solar Power Plant: Types, models, price and complete details in India 2021 (2021), https://kenbrooksolar.com/solar-power-plants/mw-solar-power-grid.

Rachit Srivastava, Solar power current status, *Challenges and Policies in India, Research & Reviews: Journal of Engineering and Technology* (2016), Vol. 5, ISSN: 2319-9873.

Yuqi Tao and Yuguo Chen, Distributed PV power forecasting using genetic algorithm based neural network approach. In *Proceedings on the International Conference on Advanced Mechatronic Systems* (2014), doi:10.1109/icamechs.2014.6911608.

Huaizhi Wang, Zhenxing Lei, Xian Zhang, Bin Zhou and Jianchun Peng, A review of deep learning for renewable energy forecasting, *Energy Conservation and Management* (2019), Vol. 198, doi:10.1016/j.enconman.2019.111799.

Harendra Kumar Yadav, Yash Pal and Madan Mohan Tripathi, A novel GA-ANFIS hybrid model for short-term solar PV power forecasting in Indian electricity market, *International Journal of Information & Optimization Sciences* (2019), Vol. 40, No. 2, pp. 377–395.

7 Deep Learning–Based Predictive Maintenance of Photovoltaic Panels

T. Yuvaraj
Saveetha Institute of Medical and Technical Sciences, Saveetha University, Chennai, India

CONTENTS

LEARNING OUTCOMES

i. To study the need and challenges for photovoltaic panel maintenance using deep learning
ii. To know the various photovoltaic maintenance methods
iii. To learn the advantages, opportunities, recommendations, and future direction of predictive maintenance of PV panel

7.1 INTRODUCTION

The high accessibility of information in the energy area makes it a favorable climate for deep learning and data science arrangements. Power sectors, energy organizations, buyers, shrewd homes, and apparatuses are some of the instances of rich information sources. They empower energy suppliers to comprehend their part in the

DOI: 10.1201/9781003202288-7

energy biological system and streamline their operations. Decreasing costs of photovoltaic (PV) frameworks, alongside improved cell efficiencies and lower instances electrical transformation faults, have prompted solar energy power generation to become a more attractive investment. As a result, the issues identified with PV systems maintenance are met with more interest, as demonstrated by the investigations and endeavors (led by different organizations and organizations) that target growing "accepted procedures" for PV framework activities. Support incorporates different exercises that are arranged into an observing system that can go from minor checks (e.g., checking the complete power produced as detailed by the inverter once each year) to high-precision observation that permits the producer or the proprietor to recognize the issues or the need for cleaning. Timely maintenance positively influences the installation's productivity and, consequently, its profitability (De Benedetti, Leonardi, Messina, Santoro, & Vasilakos, 2018).

PV power plants are growing in number and size, and this subsequently raises the significance of predictive maintenance. Optimum power generation requires observing every individual PV panel. An average checking framework comprises sensors associated with each PV panel estimating the electrical power generation of each panel as the capacity of time—for example, the power curvature. Power estimations are gathered by a wireless sensor network (WSN) and investigated utilizing information-mining apparatuses. During the preceding decase, PV checking and business frameworks have been developed and refined (Ranhotigamage & Mukhopadhyay, 2011; Papageorgas et al., 2013). A definitive objective of predictive maintenance is to recognize failing PV panels. Be that as it may, the current checking frameworks center only on gathering and envisioning operational information. We present a calculation that distinguishes a failing PV panel dependent on the power estimation history information (time arrangement of the force estimations) of the objective board and the adjoining boards (Guerriero, Di Napoli, Vallone, d'Alessandro, & Daliento, 2015; Huuhtanen & Jung, 2018).

The issue is testing a result of the enormous unique variety in the force estimations of working PV panels. These varieties occur throughout a normal operating cycle of a plant, and are a result of the following:

- The power produced at a specific time is directly related to the solar radiance received by the panel. The radiance shifts because with changes in seasons and times of day, and also depends on the geological area and exact positioning of the panel. These variables can, in practical terms, be determined precisely. Furthermore, sensors and control boards associated with these variables are usually located within close proximity to one another.
- Local weather conditions (e.g., cloud cover) impact the amount of power produced by the PV panel. Power output can drop by as much as 80% on an overcast day, compared to a sunny one. Despite having improved significantly over time, climate forecasting cannot generate completely accurate weather predictions, so even on a same day, variations in power output from a single PV panel may fluctuate significantly (Antonanzas et al. 2016). Nearby plant life and structures may also create shadow spots and cause

contrasts in solar energy received by adjoining cells. Variations may be even more pronounced in areas where topography creates long-term shadows that block sunlight from the panels at specific times of day (Lappalainen & Valkealahti 2017).
- Malfunctioning PV panels may produce progressive or quick drops in produced power.

7.2 PV PANEL MAINTENANCE

7.2.1 Need for PV Panel Maintenance

With the solar power generation inherently suitable for microgrid supply, factories and businesses are turning to solar arrays and modern PV frameworks to sidestep blackout issues related to grid supply. As a result, PV-based power panels are quickly turning into a standard instead of an alternative in India. This is further complimented by the presence of open-access networks in many states across the country. Most solar-powered panel operation and maintenance (O&M) measures require intermittent checks for guaranteeing ideal execution and security. The recurrence and level of sun-based cluster checking and sun-based board support are dependent on the establishment type, framework arrangement, and area of the business for which sunlight-based force framework is being used. Yearly assessments are suggested by organizations giving sun-powered upkeep in India. Aside from reviewing PV systems routinely, it is additionally imperative to complete assessments when their ownership changes. Solar-based PV system support measures are best completed by qualified installers or administration experts who are familiar with PV segments and their security methodology.

Effectiveness assumes a significant part in the activity of a nearby solar plant. All things considered, just introducing the PV framework is in no way, shape, or form the end of the story: ordinary support is fundamental to guarantee ideal yield from the individual PV panel. Even though PV panels are by and large low-maintenance and can work for quite a long time without issues, they are still the electrical networks. This implies that there are legally imposed standard assessments that must be seen to by PV panel operators.

7.2.2 Various PV Maintenance Methods

For this reason, nearby PV panel operators can commission a specialist organization to complete this support as framework checks and, if important, refitting or repowering measures. The master group from the SENS Service division spends significant time in the streamlining, checking, and fixing of PV panels.

As a rule, a distinction can be made between three kinds of maintenance (see also Figure 7.1).

- Corrective maintenance
- Preventive maintenance
- Predictive maintenance

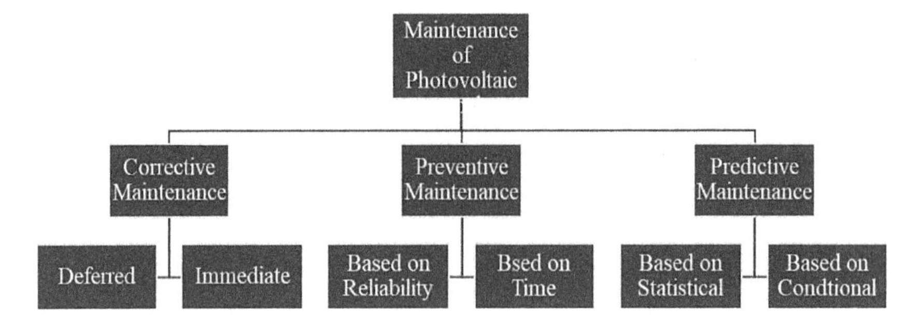

FIGURE 7.1 Various photovoltaic panel maintenance.

Corrective maintenance includes fixing the framework when an error message is received. While predictive maintenance is a possibility for the future, since it includes intercession dependent on chronicled and continuous support data, ***preventive maintenance*** is being utilized today. The advantage here is that support is not provided only in case an issue arises, but happens on a regular basis to guarantee the ideal functioning of the system. This avoids breakdowns and hence creates cost savings.

Predictive maintenance is exactly what it seems like: a support plan intended to guarantee all system parts are working as they ought to be and to keep any practical issues from decaying the system. PV panels ought to go through a deterrent support assessment at least once every year, or more frequently depending on framework size.

7.3 PREDICTIVE MAINTENANCE

7.3.1 Necessity

Recently, predictive of PV is attaining much response from the PV panel operators due to lower maintenance costs (Bosman, Leon-Salas, Hutzel, & Soto, 2020). One study reports,

> The interest for modest sustainable energy resources is at its most noteworthy that it has at any opinion been. Consequently, PV panels establishments have been being raised globally. In this challenge for sustainable energy, there is a developing requirement for exact information screening everywhere the PV based shows essential support. The trouble of finding issues and potential issues that emerge in these establishments is stupendous.
>
> (Denio, 2012)

Grimmacia and associates support Denio's assertions and feature the requirement for checking administrations by stating that:

> In regards to PV plants' dispersion, they will increment throughout the next few decades, in this manner activity and upkeep of PV panels become fundamental. The most significant level of execution will be acquired using solid and successful observing administrations with sensible expense.
>
> (Grimaccia, Aghaei, Mussetta, Leva, & Quater, 2015)

Certainly, specialists who are learning PV maintenance from an imperfection viewpoint are focusing on the requirement for preventive support and maintenance. Ancuta and Cepisca (2011) state, "We can be expected to get lifetime guarantee proficiency in assistance if the preventive maintenances are properly done." Muñoz et al. (2015) support the need for preventive maintenance: "Huge inconsistencies among the energy estimations can fill in as a caution sign to actuate upkeep activities."

A few researchers have directed studies to figure out the most worry-worthy issues for PV panels. Ancuta and Cepisca (2011) state,

> Common photovoltaic panels experience the ill effects of various issues that keep them from understanding their actual potential. A large number of the current issues originate from power faults—regardless of whether because of module bungle, direction confounds, ground shortcomings, or halfway concealing. Different issues come from framework plan limits and limitations, absence of observing, or absence of investigation capacities. Also, the shortfall of wellbeing highlights presents dangers to the two laborers and upkeep faculty.

However, data related to PV panel checking are often restricted, as indicated by the researchers: "Yet systems aren't carried out which would guarantee the nature of the establishment, neither has ideal PV site and the optimal situations for their activity been guaranteed" (Díez-Mediavilla, Alonso-Tristán, Rodríguez-Amigo, García-Calderón, & Dieste-Velasco, 2012). Further,

> Information from proprietors or support administrations are extremely scant and genuine PV panels are not typically checked. Estimation systems all things considered offices just record all out creation, which is fundamental for invoicing the energy that is delivered. Sometimes, information [is] documented in the system after the change stage.
>
> (Díez-Mediavilla, Dieste-Velasco, Rodríguez-Amigo, García-Calderón, & Alonso-Tristán, 2014)

Though uncommon in the literature, not many examinations have approached to give explicit instances of PV panels requiring maintenance. Mgonja and Saidi (2017) led an investigation surveying the viability of field support needs and practices corresponding to the execution of independent PV panels in open offices. The examination had 54 respondents, applying information assortment strategies including perceptions, meetings, and surveys. The results show that over 40% of the respondents encountered execution problems with the PV panel inside the previous 6 years.

Denio (2012) conducted a survey of the capacity for electromagnetic thermography to identify issues in PV panels utilizing UAVs. Denio's work described several cases where UAVs identified problem areas associated with warming of the environment and broken connections between cell clusters.

Jordan and associates (2015) conducted a long-term (20 years) investigation of a translucent silicon PV panel. The most essential issue identified was the need to replace the inverter multiple times over the period under study. Note that this examination took place in a lab climate with repetitive information checking utilized to recognize execution problems. In one case, the checking framework recognized combiner box arcing, which, if it had not been discovered, could have brought about additional harm to the PV systems.

7.3.2 Current Approaches and Opportunities

With the growing volume of installed PV panel capacity, the demand for testing and approving quality control keeps rising (Buerhop-Lutz & Scheuerpflug, 2015). There are a few known methodologies for evaluating PV systems for potential execution corruption (Fezzani et al., 2017):

- Visual examination of the framework and individual parts
- I-V curve and protection opposition examination of PV panels
- Infrared thermography of PV arrays
- Calculations contrasting assessed with real generation.

The first three can be summed up as manual diagnostics. The last can be separated into disappointment modes and impacts investigation approach, AI and estimating, and real-time sensor examination. Figure 7.2 gives a rundown of the methodologies for PV predictive maintenance, containing a correlation of the expense and location precision of all four methodologies. The manual analytic is the most economical yet offers the least identification precision. AI and FMEA approaches are relatively costly and provide a normal measure of location precision. Real-time sensors are the costliest but offer the highest location precision.

7.3.3 Comparison and Suitability of Deep Learning in Predictive Maintenance

Physical and information-based models for predictive maintenance (PdM) were broadly utilized 15 years prior but are less common these days because of the lack

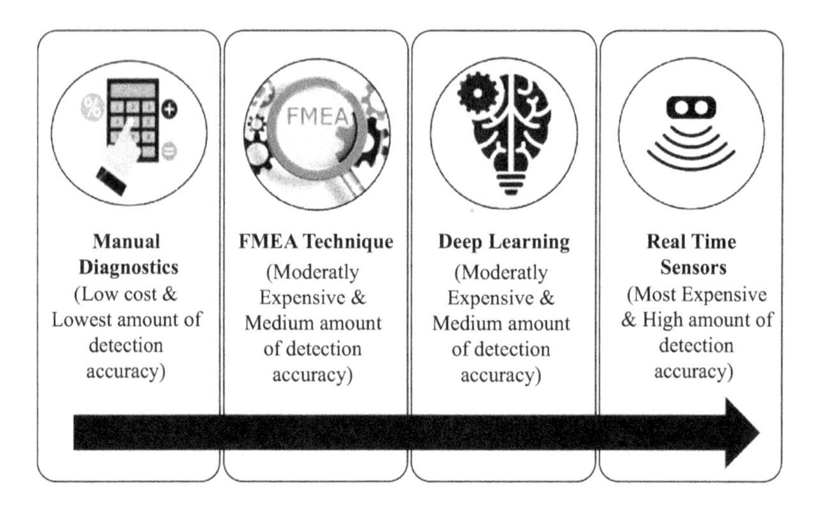

FIGURE 7.2 Summary of recent methods for PV predictive maintenance.

of capability to display complex networks. Information-driven factual and AI distributions began to grow more common in this field because they take in framework's conduct from the information directly and consequently require little area information. Most recently, because of the development of I4.0, the augmentation of computational force, and the automatizing of machine and resource information assortment, the information-driven distribution pattern has moved toward profound learning-based plans. They typically are more accurate than conventional information-driven procedures. They can operate effectively without master-level data due to their ability to separate programmed highlights related strictly to the issue being tended to. Furthermore, they can show time arrangement information utilizing consideration or time setting. The utilization of DL models is likewise broadly explored in different fields, such as picture preparing and seq2seq. In any case, their two significant downsides are the need for extensive prep data and trouble establishing model reasonableness. On the other hand, these models should be adjusted for modern and PdM information qualities and prerequisites. Thus, choosing a model for PdM application ought to be done cautiously, in the wake of examining each utilization case's necessities. Perhaps its prerequisites are not fulfilled by that second AI research pattern, which is as of now profound learning, and other types of models are more proper. For example, measurable, AI, and profound learning models have their eccentricities. They are generally ready to satisfy the accompanying PdM's qualities from the rundown by making explicit models: fast discovery and conclusion, insolubility, oddity and fault recognition, characterization mistake assessment, flexibility, and continuous calculation. The principle differences among these sorts of models are summed up in Table 7.1. Thus, choosing the best design requires an intensive investigation and correlation with deciding the one that suits both use case and its information necessities.

7.3.4 Predictive Maintenance Process Using Deep Learning

Deep learning (DL) is a subfield of AI that utilizes progressive layers of the progressively significant portrayal of information. DL incorporates various non-straight secret layers between an information layer and a yield layer. Incorporating a large number of layers and boundaries allows DL to learn extremely complex connections between data and yield. The layers, which are stacked consistently, are what we call neural organizations, and data is transferred into these layers as loads. The DL model

TABLE 7.1
Comparison of Current Approaches for Predictive Maintenance

Characteristic	Manual Diagnostics	FMEA Technique	DL
Training data Expanse	Lesser	Average	Higher
Working out time	Lesser	Average	Higher
Difficulty	Lesser	Average	Higher
Description facility	Higher	Lesser	Lesser
Accurateness	Lesser	Average	Higher

attempts to track down qualities for loads of all layers in the organization. Ultimately, the DL model will accurately plan input information to related genuine classes utilizing these loads (Korkmaz, 2019).

A loss function has been utilized to track down the ideal qualities for the loads. The loss function looks at expectations of the DL model and the genuine classes and figures a distance score. This score demonstrates how well the DL model is performing. To improve the loss score, the score delivered by the loss function is utilized as feedback. This technique is dealt with by the streamlining agent, which executes the backpropagation strategy. From the start, the worth of the loads is doled out arbitrarily, which makes the fault score high, so the loads are corrected somewhat to diminish the fault score. This cycle is repeated iteratively for various occasions (for example, age) to get the loads that limit the fault work. A DL with a base fault score yields a prepared organization that produces forecasts that are really close to the genuine classes (Korkmaz, 2019). An overall architecture and work process of a DL model are presented in Figure 7.3.

PdM anticipates when system disappointment may happen and forestalls the event of disappointment by performing maintenance. It permits the maintenance recurrence to be as infrequent as absolutely necessary to forestall spontaneous preventive maintenance, without causing costs related to doing an excess of efficient preventive support (Sikorska, Hodkiewicz, & Ma 2011; Kim, An, & Choi, 2017). Specialists concur PdM could lessen maintenance costs by 10% to 40% (Manyika et al., 2015). It likewise improves accessibility, dependability, and security of basic parts. Framework structures become increasingly more unpredictable with countless parts and complex interactions between them. Wind turbines are a genuine illustration of complex frameworks. They consist of a few basic segments such as the edges, the pitch, the gearbox, and the generator, and they present noteworthy attributes such as the center tallness, rotor measurement, or evaluated power. They ordinarily run constantly, are topographically circulated, and work under strenuous conditions. Each surprising disappointment can prompt drop inaccessibility and significant financial losses. In this manner they are an ideal contender for reserve funds through PdM. As indicated by International Standard ISO 13381-1 (Sikorska et al., 2011), Figure 7.3 presents the Predictive Maintenance Roadmap using the Pyramid Chart. PdM interaction can be isolated into three distinct advances:

(i) Diagnostics
(ii) Prognostics
(iii) Decision-making

Diagnostics include fault recognition, isolation (what part is fizzled), failure mode ID (what is the reason for disappointment), and degradation level evaluation (measurement of disappointment seriousness). The prognostic task determines the remaining useful life (RUL). The RUL is the lead time to disappointment, and a decent RUL expectation precision is significant because it affects the arranging of upkeep exercises, spare parts, and operational execution. Dynamic is an interaction bringing about the choice of right support activities among a few other options. The maintenance leader should assess each activity depending on the diagnostics or prognostics results

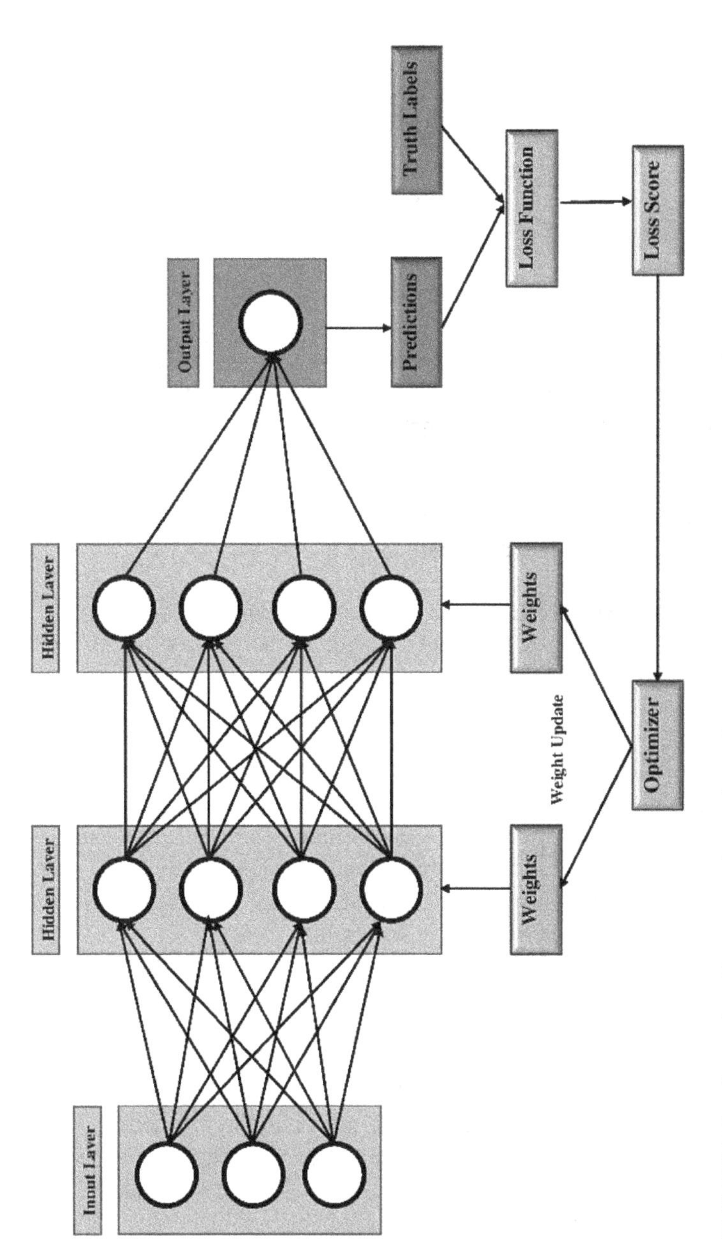

FIGURE 7.3 General architecture and workflow of a DNN model.

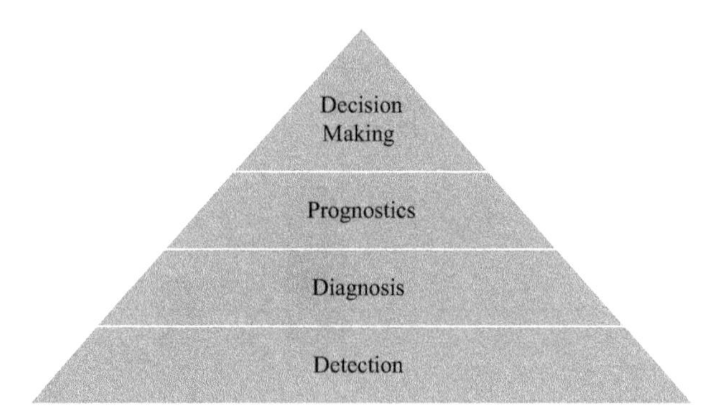

FIGURE 7.4 Representation of predictive maintenance roadmap using pyramid chart.

and ought to have the option to gauge the results of every option (Benabbou, Malki, Sankaran, & Bouzekri, 2019). For such cases SENS offers its clients block maintenance with normalized contracts. These assure straightforwardness and surrender it to the client to choose what measures are taken during maintenance. Notwithstanding the remedial and preventive administrations mentioned earlier, the particulars of such a maintenance contract incorporate master reports, system checks, and plant observing. The maintenance of a current frame is done in a few stages.

Stage 1: A system check as the reason for PV panel maintenance.

Supposed "framework checks" structure the premise of the preventive maintenance of PV panel, which expects to help ensure a fault-free yield. A framework check serves to recognize issues and enhancement potential in the current framework and to cure them directly in the short or medium term. In addition to other things, the material of the links, the trademark bends, the setting of the current controllers, the kind of inverter establishment, and the link laying nearby are recorded. For instance, flawed PV panels or inverters should be distinguished. This check empowers the current status of the framework to be resolved, and the potential for development turns out to be straightforwardly noticeable.

Stage 2: Arranging of retrofitting, change, refitting, or repowering measures.

When the framework check is completed, our specialists can assess the outcomes: along with the client, an arrangement for retrofitting, transformation, refitting, or repowering is drawn up. At this stage, producer determinations, the guidelines of the matrix administrator, the material principles, and variables such as the neighborhood climate conditions should be considered. Frequently it is adequate to supplant blemished segments if they do not, at this point, meet the most recent innovative advances, or show practical deformities.

Stage 3: Source saving acquisition of materials.

When the arranging of the concurred measures is finished, execution can start. The acquisition of materials assumes a significant part in this. SENS ensures that the necessities are resolved as decisively as could be expected, to then save assets to the greatest degree. This additionally implies extensive expense investment funds for the client.

Stage 4: Refitting and repowering the solar PV panel.

After every one of the fundamental conditions for upkeep has been explained, the refitting or repowering of the PV panel can start. Both these specialized terms are not secured and portray an updating of the whole framework or possibly singular framework segments. For the specialists at SENS, refitting implies the improvement of parts not applicable to creation, like the distant checking framework. As opposed to repowering, the impact here is regularly thought little of. Be that as it may, a difference in observing suppliers with the conceivable substitution of the information logging equipment can likewise make a critical commitment to improving response times in case of a flaw, and thus consequently increment the yield of the establishment. Because of specialized improvements specifically, there is regularly a requirement for a framework redesign.

At SENS, repowering incorporates the modernization or extension of the current PV panel, which essentially influences the power-producing parts like the PV panels and inverters. The point is to build the productivity of the framework once more, considering specifically the leftover help life, the significant feed-in duties, and the extra respect to be accomplished. These elements are a marker of whether repowering a nearby planetary group is advantageous regarding proficiency.

7.3.5 Hazards in Predictive Maintenance

Kamenopoulos and Tsoutsos (2015) expect to distinguish and assess the dangers that may emerge throughout the activity and support of PV panels. Aimed at the improvement of the examination, the procedure utilized was an operative hazard board, which consists of issues related to operation and support, and whose system of activity incorporates the accompanying advances:

- Identify wellsprings of risk
- Assess risks
- Make hazard choices
- Instrument controls
- Modification of monitor and control.

The initial step has been to decide the cycles and methods to be utilized to survey the weight given to each chance by making a work network where the likelihood of achievement of hazard is identified with the seriousness of the danger by allocating a worth of hazard level to each conceivable occasion. The degrees of hazard utilized are:

1. Negligible
2. Minor
3. Moderate

4. Serious
5. Critical.

The scope of danger is assessed by four distinct degrees: (1) catastrophic (it can cause loss of the facility), (2) critical (it can cause significant damage to the facility and/or serious injuries or illnesses to personnel), (3) marginal (it can cause minor damage to the facility and/or minor wounds to personnel), and (4) negligible (it doesn't detrimentally affect the well-being of the facility or its personnel). Additionally, there are three classifications of hazard recognizable proof: peril source (normal, human, or innovative risks), the impact of danger (on laborers, dangers to outsiders, or perils to sun-facing boards), and the interaction of activities (activity or upkeep of offices). When potential dangers are distinguished, a specialist in the activity and upkeep of photovoltaic frameworks is approached to assess the likelihood that the dangerous event will occur and the extent of consequences (following the measures checked previously). Likewise, a weight is allocated to each chance with a worth of 0–1 to weigh the outcomes.

7.3.6 Advantages of Preventive Maintenance

Preventive maintenance is fundamental to extending the lifecycle of solar PV power plants. Consequently, insightful checking frameworks have been introduced to distinguish potential breakdowns through the appraisal of PV plants exhibitions.

A monitoring solution like Track So helps in executing maintenance by monitoring solar PV sources, as it provides real-time data, error and anomaly detection, and visual analysis of data through charts and graphs. This helps in maintaining the health and conditioning of your PV sources as well as helps take measures required for optimum output. All the real-time data collected along with the historical data can be easily extrapolated and used for the prediction of upcoming scenarios and hence indicate problems even before they occur.

Predictive maintenance has the following benefits:

- Reduced segment fix and substitution costs
- Reduced income loss resulting from downtime
- Focused O&M action
- Better stock administration
- Improved part life span inferable from better operational practices

In the present power economy where the presentation of environmentally friendly power resources is critical to its venture practicality and future speculation, preventive support is the best approach.

7.3.7 Recommendations and Future Direction

- With PV-based establishments, innovation is powering an advancement from responsive to predictive maintenance. Artificial intelligence (AI), associated sensors and information, and examination can help forestall exorbitant vacation and assist representatives with staying away from risky crisis fix work.

- As PV-based energy becomes more and more competitive with nonrenewable fuel-based sources in terms of cost and operational efficiency, advanced tech can be the next best step. A mechanized robot armada can reduce examination time and human costs. Distant detecting, controlled by the Industrial Internet of Things (IIoT), can trigger preventive maintenance exercises to expand the future of stuff boxes, direction, and other gear.
- With an interface on a PC or a smartphone, far-off sensors can give warnings continuously, possibly foreseeing issues ahead of their arising. Technicians can proactively arrange parts, plan downtime for critical support, and report real conditions to increase hardware longevity. Over the long run, AI can screen and search for examples and patterns in the information to better foresee and conceivably improve future maintenance and operations.
- The capacity to adopt an all-encompassing strategy for assessing information across all resources and areas is possibly the greatest benefit for PV panel's establishments.
- "With customary SCADA (Supervisory Control and Data Acquisition) frameworks, there were limits on the capacity to decipher the information," said Bill Barbato, a Travelers Risk Control specialist. Therefore, the framework warnings demonstrated the need for restorative activity instead of alerting administrators before damage ocurred. "Presently, with AI, more information can prompt better investigation and more prominent precision. The frameworks can perceive designs, with contribution from administrators to help manage to learn."
- Installers can lead distant nonstop checking, including oil inspecting, quality observing, and far-off vibration checking. Interfacing gadgets consider gathering and breaking down information to possibly distinguish patterns. By applying AI, it is feasible to create preventive models that can diminish the potential for unanticipated faults and ultimately reduce operational costs.
- It is imperative to keep in mind that innovation does not replace the need for customary support programs at solar power establishments. Probably the best upkeep programs we've seen include the establishment proprietor utilizing prescient support to expect issues and help set up the recurrence of that maintenance.
- There is the potential danger that operators and maintenance groups will experience the ill effects of "alarm weariness" if frameworks give them constantly. Having approaches set up concerning how representatives react to alerts can help relieve these dangers.
- Further work to improve performance tracking should incorporate a more extensive arrangement of meetings of industry entrepreneurs, along with technology, HR, and financial specialists, to achieve a more extensive agreement and alternate points of view from an organizational perspective on how these measurements are impacted by the utilization of AI.

7.4 CONCLUSION

There are numerous things that can go wrong during operation of a solar power plant. Electric faults and structural damage that go unnoticed and/or unrectified for prolonged periods of time lead to diminished power output. Even a single malfunctioning PV cell within a single microgrid board can cause operational issues that, if left unresolved, may cascade into bigger, more complex problems. Without having a tracking and analyzing approach, combined with a solid, practical maintenance plan, it is impossible to distinguish, evaluate, and resolve these problems. Considering the scope of the issue and gaps withing present arrangements, we suggest that PV panel proprietors need an off-the-rack framework-level preventive maintenance instrument to limit downtime, improve performance, and increase revenue.

A tracking and analysis system like Track So helps in executing maintenance by observing solar-based PV resources, recognizing faults, and analyzing data through outlines and diagrams. This chapter presented a brief discussion of deep learning-based predictive maintenance of PV panels. All the ongoing data gathered alongside historical data can be effectively utilized for predicting the impending faults.

REFERENCES

Ancuta, F., & Cepisca, C. (2011). Fault analysis possibilities for PV panels. In *Proceedings of the 2011 3rd International Youth Conference on Energetics (IYCE)* (pp. 1–5). IEEE.

Antonanzas, J., Osorio, N., Escobar, R., Urraca, R., Martinez-de-Pison, F. J., & Antonanzas-Torres, F. (2016). Review of photovoltaic power forecasting. *Solar Energy*, *136*, 78–111.

Benabbou, L., Malki, Z., Sankaran, K., & Bouzekri, H. (2019). Machine learning-based predictive maintenance for renewable energy: The case of power plants in Morocco. In *Proceedings of the 36 International Conference on Maching Learning*. Long Beach, CA, USA.

Bosman, L. B., Leon-Salas, W. D., Hutzel, W., & Soto, E. A. (2020). PV system predictive maintenance: Challenges, current approaches, and opportunities. *Energies*, *13*(6), 1398.

Buerhop-Lutz, C., & Scheuerpflug, H. (2015). Inspecting PV-plants using aerial, drone-mounted infrared thermography system. In *3rd Southern African Solar Energy Conference*. South Africa, 11–13 May, 2015.

De Benedetti, M., Leonardi, F., Messina, F., Santoro, C., & Vasilakos, A. (2018). Anomaly detection and predictive maintenance for photovoltaic systems. *Neurocomputing*, *310*, 59–68.

Denio, H. (2012). Aerial solar thermography and condition monitoring of photovoltaic systems. In *2012 38th IEEE Photovoltaic Specialists Conference* (pp. 000613–000618). IEEE.

Díez-Mediavilla, M., Alonso-Tristán, C., Rodríguez-Amigo, M. D. C., García-Calderón, T., & Dieste-Velasco, M. I. (2012). Performance analysis of PV plants: Optimization for improving profitability. *Energy Conversion and Management*, *54*(1), 17–23.

Díez-Mediavilla, M., Dieste-Velasco, M. I., Rodríguez-Amigo, M. D. C., García-Calderón, T., & Alonso-Tristán, C. (2014). Performance of grid-tied PV facilities based on real data in Spain: Central inverter versus string system. *Energy Conversion and Management*, *86*, 1128–1133.

Fezzani, A., Mahammed, I. H., Drid, S., Zaghba, L., Bouchakour, A., & Benbitour, M. K. (2017). Experimental investigation of effects of partial shading and faults on photovoltaic modules performances. *Leonardo Electronic Journal of Practices and Technologies*, *11*, 183–200.

Grimaccia, F., Aghaei, M., Mussetta, M., Leva, S., & Quater, P. B. (2015). Planning for PV plant performance monitoring by means of unmanned aerial systems (UAS). *International Journal of Energy and Environmental Engineering*, *6*(1), 47–54.

Guerriero, P., Di Napoli, F., Vallone, G., d'Alessandro, V., & Daliento, S. (2015). Monitoring and diagnostics of PV plants by a wireless self-powered sensor for individual panels. *IEEE Journal of Photovoltaics*, *6*(1), 286–294.

Huuhtanen, T., & Jung, A. (2018). Predictive maintenance of photovoltaic panels via deep learning. In *2018 IEEE Data Science Workshop (DSW)* (pp. 66–70). IEEE.

Jordan, D. C., Sekulic, B., Marion, B., & Kurtz, S. R. (2015). Performance and aging of a 20-year-old silicon PV system. *IEEE Journal of Photovoltaics*, *5*(3), 744–751.

Kamenopoulos, S. N., & Tsoutsos, T. (2015). Assessment of the safe operation and maintenance of photovoltaic systems. *Energy*, *93*, 1633–1638.

Kim, N. H., An, D., & Choi, J. H. (2017). *Prognostics and health management of engineering systems*. Switzerland: Springer International Publishing.

Korkmaz, S. (2019). Small drug molecule classification using deep neural networks. *Turkiye Klinikleri Journal of Biostatistics*, *11*(2), 1–9.

Lappalainen, K., & Valkealahti, S. (2017). Photovoltaic mismatch losses caused by moving clouds. *Solar Energy*, *158*, 455–461.

Manyika, J., Chui, M., Bisson, P., Woetzel, J., Dobbs, R., Bughin, J., & Aharon, D. (2015). *The Internet of things: Mapping the value beyond the hype*. New York: McKinsey & Co.

Mgonja, C. T., & Saidi, H. (2017). Effectiveness on implementation of maintenance management system For off-grid solar PV systems In public facilities–a case study of SSMP1 project In Tanzania. *International Journal of Mechanical Engineering and Technology*, 8(7), 869–880.

Muñoz, C. Q. G., Marquez, F. P. G., Liang, C., Maria, K., Abbas, M., & Mayorkinos, P. (2015). A new condition monitoring approach for maintenance management in concentrate solar plants. In *Proceedings of the Ninth International Conference on Management Science and Engineering Management* (pp. 999–1008). Springer, Berlin, Heidelberg.

Papageorgas, P., Piromalis, D., Antonakoglou, K., Vokas, G., Tseles, D., & Arvanitis, K. G. (2013). Smart solar panels: In-situ monitoring of photovoltaic panels based on wired and wireless sensor networks. *Energy Procedia*, *36*, 535–545.

Ranhotigamage, C., & Mukhopadhyay, S. C. (2011). Field trials and performance monitoring of distributed solar panels using a low-cost wireless sensors network for domestic applications. *IEEE Sensors Journal*, *11*(10), 2583–2590.

Sikorska, J. Z., Hodkiewicz, M., & Ma, L. (2011). Prognostic modelling options for remaining useful life estimation by industry. *Mechanical Systems and Signal Processing*, *25*(5), 1803–1836.

Index

R

S

T

U

V

W

Z